ALSO BY HENRY PETROSKI

The Pencil

Beyond Engineering

To Engineer Is Human

THE EVOLUTION OF USEFUL THINGS

THE EVOLUTION OF USEFUL THINGS

Henry Petroski

Alfred A. Knopf New York 1992

THIS IS A BORZOI BOOK
PUBLISHED BY ALFRED A. KNOPF, INC.

Some of this material first appeared, in a different form, in *American Heritage of Invention and Technology, Technology Review, Wigwag,* and *Wilson Quarterly.*

Grateful acknowledgment is made to the following for permission to reprint previously published material:

Barrie & Jenkins: Excerpts from *The Nature and Aesthetics of Design* by David Pye (Barrie & Jenkins, London, 1978). Reprinted by permission of Barrie & Jenkins, a division of The Random Century Group Limited.

Caliban Books: Excerpt from *William Smith, Potter and Farmer, 1790–1858* by George Sturt (Caliban Books, London, 1978). Reprinted by permission.

HarperCollins Publishers Inc.: Excerpts from *Etiquette: The Blue Book of Social Usage* by Emily Post. Copyright © 1965 by Funk and Wagnalls Co., Inc. Reprinted by permission of HarperCollins Publishers Inc.

Harvard University Press: Excerpts from *Notes on the Synthesis of Form* by Christopher Alexander (Harvard University Press, 1964). Reprinted by permission.

Library of Congress Cataloging-in-Publication Data

Petroski, Henry.
The evolution of useful things/by Henry Petroski.
—1st ed.
p. cm.
Includes bibliographical references and index.
ISBN 0-679-41226-3
1. Inventions. 2. Patents. I. Title.
T212.P465 1992
609—dc20 91-39524
CIP

Manufactured in the United States of America

First Edition

to my mother,

and to the memory of my father

Contents

Preface

Other than the sky and some trees, everything I can see from where I now sit is artificial. The desk, books, and computer before me; the chair, rug, and door behind me; the lamp, ceiling, and roof above me; the roads, cars, and buildings outside my window, all have been made by disassembling and reassembling parts of nature. If truth be told, even the sky has been colored by pollution, and the stand of trees has been oddly shaped to conform to the space allotted by development. Virtually all urban sensual experience has been touched by human hands, and thus the vast majority of us experience the physical world, at least, as filtered through the process of design.

Given that so much of our perception involves made things, it is reasonable to ask how they got to look the way they do. How is it that an artifact of technology has one shape rather than another? By what process do the unique, and not-so-unique, designs of manufactured goods come to be? Is there a single mechanism whereby the tools of different cultures evolve into distinct forms and yet serve the same essential function? To be specific, can the development of the knife and fork of the West be explained by the same principle that explains the chopsticks of the East? Can any single theory explain the shape of a Western saw, which cuts on the push stroke, as readily as an Eastern one, which cuts on the pull? If form does not follow function in any deterministic way, then by what mechanism do the shapes and forms of our made world come to be?

Such are the questions that have led to this book. It extends an exploration of engineering that I began in *To Engineer Is Human,* which dealt mainly with understanding why made things break, and that I continued in *The Pencil,* which traced the evolution of a single artifact through the cultural, political, and technological vicissitudes of history. Here I have focused not on the physical failings of any single thing but, rather, on the implications of failure—whether physical, functional, cultural, or psychological—for form generally. This extended essay, which may be read as a refutation of the design dictum that "form follows function," has led to considerations that go beyond things themselves to the roots of the often ineffable creative processes of invention and design.

As artifacts evolve from artifacts, so do books from books. In writing this one, I have once again benefited from the physical and intellectual resources of many libraries and librarians. As always, I acknowledge Eric Smith, head of Duke University's Vesić Engineering Library, who remains ever-patient in the face of my frequently vague requests for often obscure sources, and even pursues avenues of information I would never have dreamed of following. Stuart Basefsky, of the Public Documents Department in Duke's Perkins Library, helped me get oriented in patent literature, which proved to be so important for my case, and the patent repository of the D. C. Hill Library of North Carolina State University graciously filled my numerous requests for documents. Several manufacturers, by freely providing their company histories, catalogues, and ephemera, enabled me to read beyond library walls and find invaluable documentation of things as they have been and are. Also, many friends, readers, and collectors generously shared with me art, facts, and artifacts that have found their way into my work. Where I have remembered my debts, I have acknowledged them in the notes at the end of this volume.

Correspondence and conversations with inventors and designers over the years have certainly shaped the ideas in this book, but, as in so much invention and design, individual contributions must necessarily remain largely anonymous, because they have become so threaded into the fabric of the work that to try to pick out even the most conspicuous of them would but lead to a lot of loose ends. Where practitioners have written or spoken for the record, their works are referenced in my bibliography, as are all those in which I can recall having read support for my thesis. By their example and encouragement, certain writers, engineers, and historians of technology have been especially instrumental in influencing this book, and I must single out Freeman Dyson, Eugene Ferguson, Melvin Kranzberg, and Walter Vincenti for their support.

A book naturally takes time and space to write, and I am indebted to a fellowship from the John Simon Guggenheim Memorial Foundation for the former and to a carrel in Perkins Library for the latter.

I am grateful to my supportive editor, Ashbel Green, and to the many others at Alfred A. Knopf who have read the manuscript with pencils of various colors and in other ways prepared it for the press. Whatever shortcomings that remain are naturally my responsibility. Finally, my family once again understood my need to think and read at home each evening, and they quietly and constantly added to my store of examples by leaving interesting thing after interesting thing, from the broken to the bizarre, on my desk. I am, as always, grateful to Stephen, to Karen, who indexed this book, and especially to Catherine, who read the book for me at each stage of its evolution.

William R. Perkins Library
Duke University
April 1992

THE EVOLUTION OF USEFUL THINGS

1

How the Fork Got Its Tines

The eating utensils that we use daily are as familiar to us as our own hands. We manipulate knife, fork, and spoon as automatically as we do our fingers, and we seem to become conscious of our silverware only when right- and left-handers cross elbows at a dinner party. But how did these convenient implements come to be, and why are they now so second-nature to us? Did they appear in some flash of genius to one of our ancestors, who yelled "Eureka!," or did they evolve as naturally and quietly as did the parts of our bodies? Why is Western tableware so alien to Eastern cultures, and why do chopsticks make our hands all thumbs? Are our eating utensils really "perfected," or is there room for improvement?

Such questions that arise out of table talk can serve as paradigms for questions about the origins and evolution of all made things. And seeking answers can provide insight into the nature of technological development generally, for the forces that have shaped place settings are the same that have shaped all artifacts. Understanding the origins of diversity in pieces of silverware makes it easier to understand the diversity of everything from bottles, hammers, and paper clips to bridges, automobiles, and nuclear-power plants. Delving into the evolution of the knife, fork, and spoon can lead us to a theory of how all the things of technology evolve. Exploring the tableware that we use every day, and yet know so little about, provides as good a starting point for a consideration of the interrelated natures of invention, innovation, design, and engineering as we are likely to find.

Some writers have been quite unequivocal about the origins of things. In their *Picture History of Inventions,* Umberto Eco and

G. B. Zorzoli state flatly that "all the tools we use today are based on things made in the dawn of prehistory." And in his *Evolution of Technology,* George Basalla posits as fundamental that "any new thing that appears in the made world is based on some object already there." Such assertions appear to be borne out in the case of eating utensils.

Certainly our earliest ancestors ate food, and it is reasonable to ask how they ate it. At first, no doubt, they were animals as far as their table manners were concerned, and so we can assume that the way we see real animals eat today gives us clues as to how the earliest people ate. They would use their teeth and nails to tear off pieces of fruits, vegetables, fish, and meat. But teeth and nails can only do so much; they alone are generally not strong enough or sharp enough to render easily all things edible into bite-sized pieces.

The knife is thought to have had its origins in shaped pieces of flint and obsidian, very hard stone and rock whose fractured edges can be extremely sharp and thus suitable to scrape, pierce, and cut such things as vegetable and animal flesh. How the efficacious properties of flints were first discovered is open to speculation, but it is easy to imagine how naturally fractured specimens may have been noticed by early men and women to be capable of doing things their hands and fingers could not. Such a discovery could have occurred, for example, to someone walking barefoot over a field and cutting a foot on a shard of flint. Once the connection between accident and intention was made, it would have been a matter of lesser innovation to look for other sharp pieces of flint. Failing to find an abundance of them, early innovators might have engaged in the rudiments of knapping, perhaps after noticing the naturally occurring fracture of falling rocks.

In time, prehistoric people must have come to be adept at finding, making, and using flint knives, and they would naturally also have discovered and developed other ingenious devices. With fire came the ability to cook food, but even meat that had been delicately cut into small pieces could barely be held over a fire long enough to warm it, let alone cook it, and sticks may have come to be used in much the same way as children today roast marshmallows. Pointed sticks, easily obtained in abundance from nearby trees and bushes, could have been used to keep an individual's fingers from being cooked with dinner. But larger pieces of meat, if not the whole animal, would more likely first have been roasted on a larger stick.

This damascened blade of a thousand-year-old Saxon scramasax is inscribed, "Gebereht owns me." Early knives were proud personal possessions and they served many functions; the pointed blade not only could pierce the flesh of an enemy but also could spear pieces of food and convey them to the mouth. This knife's long-missing handle may have been made of wood or bone.

Upon being removed from the fire, the roast could be divided among the diners, perhaps by being scored first with a flint knife. Those around the fire could then pick warm pieces of tender meat off the bone with pointed sticks, or resort to their fingers.

From the separate implements of sharp-edged flint for cutting and sharp-pointed stick for spearing evolved the single implement of a knife that would be easily recognized as such today. By ancient times, knives were being made of bronze and iron, with handles of wood, shell, and horn. The applications of these knives were multifarious, as tools and weapons as well as dining implements, and in Saxon England a knife known as a "scramasax" was the constant companion of its owner. Whereas common folk still ate mostly with their teeth and fingers, tearing meat from the bone with abandon, more refined people came to employ their knives in some customary ways. In the politest of circumstances, the dish being sliced might have been held steady by a crust of bread, with the knife being used also to spear the morsel and convey it to the mouth, thus keeping the fingers of both hands clean.

I first experienced what it is like to eat with only a single knife some years ago in Montreal, in a setting that might best be described as participatory dinner theater. The Festin du Gouverneur took place in an old fort, and a hundred or so of us sat at long bare wooden tables set parallel to three sides of a small stage. At each place were a napkin and a single knife, with which we were expected to eat our entire meal, which consisted of roast chicken, potatoes, carrots, and a roll. It was relatively easy to deal with the firm carrots and potatoes, for pieces of them could be sheared off with the knife blade, speared on its point, and put neatly in the mouth. However,

I had considerable trouble just cutting off pieces of chicken. At first I tried to steady it with my roll, but it was soft to begin with and soon became crumbly and soggy. I had to resort to eating the chicken with my fingers. What I remember most about the experience was how greasy my fingers felt for the rest of the evening. How convenient and more civilized it would have been at least to have had a second knife.

My only other experience eating with a single knife occurred at a barbecue restaurant popular with the students and faculty of Texas A & M University. I had been visiting the campus, and for a light dinner before I caught my plane back to North Carolina one of my hosts thought I might enjoy trying what he described as real barbecue—Texas beef instead of the pork variety I had come to know and love in the Southeast. I ordered a small portion of the house specialty, and the waitress brought me several slices of beef brisket, two whole cooked onions, a fat dill pickle, a good-sized wedge of cheddar cheese, and two slices of white bread, all wrapped in a large piece of white butcher paper, which when opened up served as both plate and place mat. On the paper was set a very sharply pointed butcher knife with a bare wooden handle.

I followed the lead of the Aggies I was with and picked up a piece of brisket with the point of the knife and laid it atop a piece of bread. (In medieval times, the piece of bread, called a "trencher," would have been four days old to give it some stiffness and body, the better to hold the meat and sauce.) We proceeded to cut off bite-sized pieces of this open-faced sandwich, and everything else set before us, and it all was delicious. The single knife worked well, because it was very sharp and could be pressed through the firm food, which itself did not slip much on the paper. However, I was quite distracted throughout the meal by my host, who used his knife so casually that I feared any minute he would cut his lip or worse. He also kept me a bit uneasy with his jokingly expressed hope that no one would come up behind us and give us a good pat on the back just as we were putting our knives into our mouths.

Eating a meal with two knives might seem to have been doubly crude and dangerous, but in its time it was thought of as the height of refinement. For the most formal dining in the Middle Ages, a knife was grasped in each hand. For a right-handed person the knife in the left hand held the meat steady while the knife in the right hand sliced off an appropriately sized piece. This piece was then

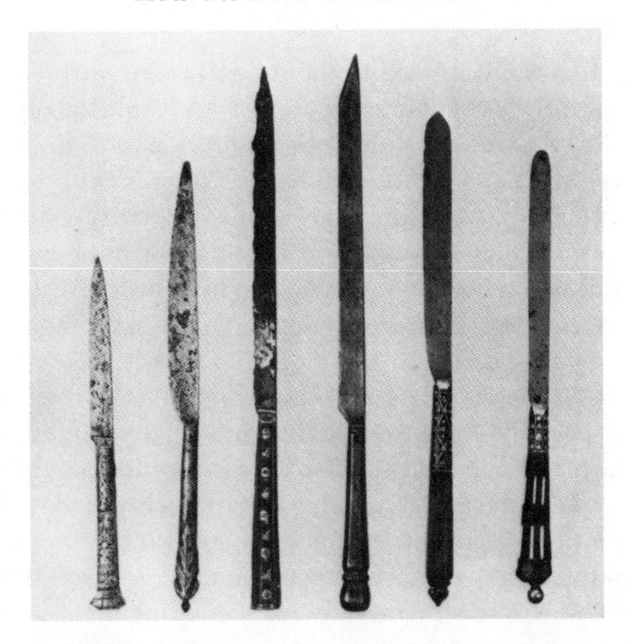

Knives, like all artifacts, have over time been subject to the vagaries of style and fashion, especially in the more decorative aspects of their handles. These English specimens date (left to right) from approximately 1530, 1530, 1580, 1580, 1630, and 1633, and they show that in one form or another the functional tip of the knife remained a constant feature until the introduction of the fork provided an alternate means of spearing food.

speared and conveyed to the mouth on the knife's tip. Eating with two knives represented a distinct advance in table manners, and the adept diner must have manipulated a pair of knives as readily as we do a knife and fork today.

By using one knife to steady the roast in the middle of the table while the other knife cut off a slice, the diners could help themselves without touching the common food. But a sharp, pointed knife is not a very good holding device, as we can easily learn by trying to eat a T-bone with a steak knife in each hand. If the holding knife is to press the steak against the plate, we must use some effort to keep it in place, and this can become tiring; if the holding knife is to spear the steak, we will soon find it rotating in place like a wheel on an axle. As a result, using the fingers to steady food being cut was not uncommon.

Frustrations with knives, especially their shortcomings in holding meat steady for cutting, led to the development of the fork. While ceremonial forks were known to the Greeks and Romans, they apparently had no names for table forks, or at least did not use them in their writings. Greek cooks did have a "flesh-fork . . . to take meat from a boiling pot," and this kitchen utensil "had a resemblance to the hand, and was used to prevent the fingers from being scalded." Ancient forklike tools also included the likes of hay forks and Neptune's trident, but forks are assumed not to have been used for dining in ancient times.

The first utilitarian food forks had two prongs or tines, and were employed principally in the kitchen and for carving and serving. Such forks pierced the meat like a pointed knife, but the presence of two tines kept the meat from moving and twisting too easily when a piece was being sliced off. Although this advantage must also have been recognized in prehistoric eras, when forked sticks were almost as easy to come by as straight ones for skewering meat over the fire, the fork as an eating utensil was a long time in coming. It is believed that forks were used for dining in the royal courts of the Middle East as early as the seventh century and reached Italy around the year 1100. However, they did not come into any significant service there until about the fourteenth century. The inventory of Charles V of France, who reigned from 1364 to 1380, listed silver and gold forks, but with an explanation "that they were only used for eating mulberries and foods likely to stain the fingers." Table forks for conveying a variety of foods to the mouth moved westward to France with Catherine de Médicis in 1533, when she married the future King Henry II, but the fork was thought to be an affectation, and those who lost half their food as it was lifted from plate to mouth were ridiculed. It took a while for the new implement to gain widespread use among the French.

Not until the seventeenth century did the fork appear in England. Thomas Coryate, an Englishman who traveled in France, Italy, Switzerland, and Germany in 1608, published three years later an account of his adventures in a book entitled, in part, *Crudities Hastily Gobbled Up in Five Months.* At that time, when a large piece of meat was set on a table in England, the diners were still expected to partake of this main dish by slicing off a portion each while holding the roast steady with the fingers of their free hand. Coryate saw it done differently in Italy:

> I observed a custom in all those Italian cities and towns through which I passed, that is not used in any other country that I saw in my travels, neither do I think that any other nation of Christendom doth use it, but only Italy. The Italian, and also most strangers that are commorant in Italy, do always at their meals use a little fork when they cut their meat. For while with their knife which they hold in one hand they cut the meat out of the dish, they fasten the fork, which they hold in their other hand, upon the same dish; so that whatsoever he be that sitting in the company of any others at the meal, should unadvisedly touch the dish of meat with his fingers from which all at the table do cut, he will give occasion of offense unto the company, as having transgressed the laws of good manners, insomuch that for his error he shall be at least brow beaten if not reprehended in words. This form of eating I understand is generally used in all places of Italy; their forks being for the most part made of iron or steel, and some of silver, but those are used only by gentlemen. The reason of this their curiosity is, because the Italian cannot by any means indure to have his dish touched with fingers, seeing all men's fingers are not alike clean. Hereupon I myself thought to imitate the Italian fashion by this forked cutting of meat, not only while I was in Italy, but also in Germany, and oftentimes in England since I came home.

Coryate was jokingly called "Furcifer," which meant literally "fork bearer," but which also meant "gallows bird," or one who deserved to be hanged. Forks spread slowly in England, for the utensil was much ridiculed as "an effeminate piece of finery," according to the historian of inventions John Beckmann. He documented further the initial reaction to the fork by quoting from a contemporary dramatist who wrote of a "fork-carving traveller" being spoken of "with much contempt." Furthermore, no less a playwright than Ben Jonson could get laughs for his characters by questioning, in *The Devil Is an Ass,* first produced in 1616,

> *The laudable use of forks,*
> *Brought into custom here as they are in Italy,*
> *To the sparing of napkins.*

But the new fashion was soon being taken more seriously, for Jonson could also write, in *Volpone,* "Then must you learn the use and handling of your silver fork at meals."

Putting aside acceptance and custom, what makes the fork work, of course, are its tines. But how many tines make the best fork, and why? Something with a single tine is hardly a fork, and would be no better than a pointed knife for spearing and holding food. The toothpicks at cocktail parties may be considered, like sharpened sticks, rudimentary forks, but most of us have experienced the frustrations of manipulating a toothpick to pick up a piece of shrimp and dip it in sauce. If the shrimp does not fall off, it rotates in the sauce cup. If the shrimp does not drop into the cup, we must contort our hand to hold the toothpick, shrimp, and dripping sauce toward the vertical while trying to put the hors d'oeuvre on our horizontal tongue. The single-tined fork is not generally an instrument of choice, but that is not to say it does not have a place. Butter picks are really single-tined forks, but, then, we do want a butter pick to release the butter easily. Escargot and nut picks might also be classified as single-pronged forks, but, then, there is hardly room for a second tine in a snail's snug spiral or a pecan shell's interstices.

The two-pronged fork is ideal for carving and serving, for a roast can be held in place without rotating, and the fork can be slid in and out of the meat relatively easily. The implement can be moved along the roast with little difficulty and can also convey slices of meat from carving to serving platter with ease. The carving fork functions as it was intended, leaving little to be desired, and so it has remained essentially unchanged since antiquity. But the same is not true of the table fork.

As the fork grew in popularity, its form evolved, for its shortcomings became evident. The earliest table forks, which were modeled after kitchen carving forks, had two straight and longish tines that had developed to serve the principal function of holding large pieces of meat. The longer the tines, the more securely something like a roast could be held, of course, but longish tines are unnecessary at the dining table. Furthermore, fashion and style dictated that tableware look different from kitchenware, and so since the seventeenth century the tines of table forks have been considerably shorter and thinner than those of carving forks.

In order to prevent the rotation of what was being held for cutting, the two tines of the fork were necessarily some distance apart, and this spacing was somewhat standardized. However, small loose pieces of food fell through the space between the tines and thus could not be picked up by the fork unless speared. Furthermore, the very advantage of two tines for carving meat, their ease of removal, made it easy for speared food to slip off early table forks. Through the introduction of a third tine, not only could the fork function more efficiently as something like a scoop to deliver food to the mouth, but also food pierced by more tines was less likely to fall off between plate and mouth.

If three tines were an improvement, then four were even better. By the early eighteenth century, in Germany, four-tined forks looked as they do today, and by the end of the nineteenth century the four-tined dinner fork became the standard in England. There have been five- and six-tined forks, but four appears to be the optimum. Four tines provide a relatively broad surface and yet do not feel too wide for the mouth. Nor does a four-tined fork have so many tines that it resembles a comb, or function like one when being pressed into a piece of meat. Wilkens, the German silversmith, does make a modern five-tined dinner fork, but it appears to have been designed more for fashion than function, since the pattern (called Epoca) is marketed as being "unique in its entirety and in every detail" and "full of generous, massive strength." The fork's selling point seems to be its unusual appearance rather than its effectiveness for eating. Many contemporary silverware patterns have three-tined dinner forks for similar reasons, but some go so far in rounding and tapering the tines, thus softening the lines of the fork, that it is almost impossible to pick up food with it.

The evolution of the fork in turn had a profound impact on the evolution of the table knife. With the introduction of the fork as a more efficient spearer of food, the pointed knife tip became unnecessary. But many artifacts retain nonfunctional vestiges of earlier forms, and so why did not the knife? The reason appears to be at least as much social as technical. When everyone carried a personal knife not only as a singular eating utensil but also as a tool and a defensive weapon, the point had a purpose well beyond the spearing of food. Indeed, many a knife carrier may have preferred to employ his fingers for lifting food to his mouth rather than the tip of his most prized possession. According to Erasmus's 1530 book on manners, it was not impolite to resort to fingers to help yourself from the pot as long as you "use only three fingers at most" and you "take the first piece of meat or fish that you touch." As for the knife, the young were admonished, "Don't clean your teeth with your knife." A French book of advice to students recognized the implicit threat involved in using a weapon at the table, and instructed its readers to place the sharp edge of their knife facing toward themselves, not their neighbor, and to hold it by its point in passing it to someone else. Such customs have influenced how today's table is set and how we are expected to behave at it. In Italy, for example, when one is eating with a fork alone, it is correct to rest the free hand in full view on the table edge. Though this might be considered poor manners in America, the custom is believed to have originated in the days when the visible hand showed one's fellow diners that no weapon was being held in the lap.

It is said to have been Cardinal Richelieu's disgust with a frequent dinner guest's habit of picking his teeth with the pointed end of his knife that drove the prelate to order all the points of his table knives ground down. In 1669, as a measure to reduce violence, King Louis XIV made pointed knives illegal, whether at the table or on the street. Such actions, coupled with the growing widespread use of forks, gave the table knife its now familiar blunt-tipped blade. Toward the end of the seventeenth century, the blade curved into a scimitar shape, but this contour was to be modified over the next century to become less weaponlike. The blunt end became more prominent, not merely to emphasize its bluntness but, since the paired fork was likely to be two-tined and so not an efficient scoop, to serve as a surface onto which food might be heaped for conveying to the mouth. Peas and other small discrete foods, which had been eaten by being pierced one by one with a knife point or a fork tine, could now be eaten more efficiently by being piled on the knife blade, whose increasingly backward curve made it possible to insert the food-laden tip into the mouth with less contortion of the wrist. During this time, the handles on some knife-and-fork sets became pistol-shaped, thus complementing the curve of the knife blade but making the fork look curiously asymmetrical.

With the beginning of the nineteenth century, English table-knife blades came to be made with nearly parallel straight sides, perhaps in part as a consequence of the introduction of steam power during the Industrial Revolution and the economy of process in forming this shape out of ingots, but perhaps even more because the fork had

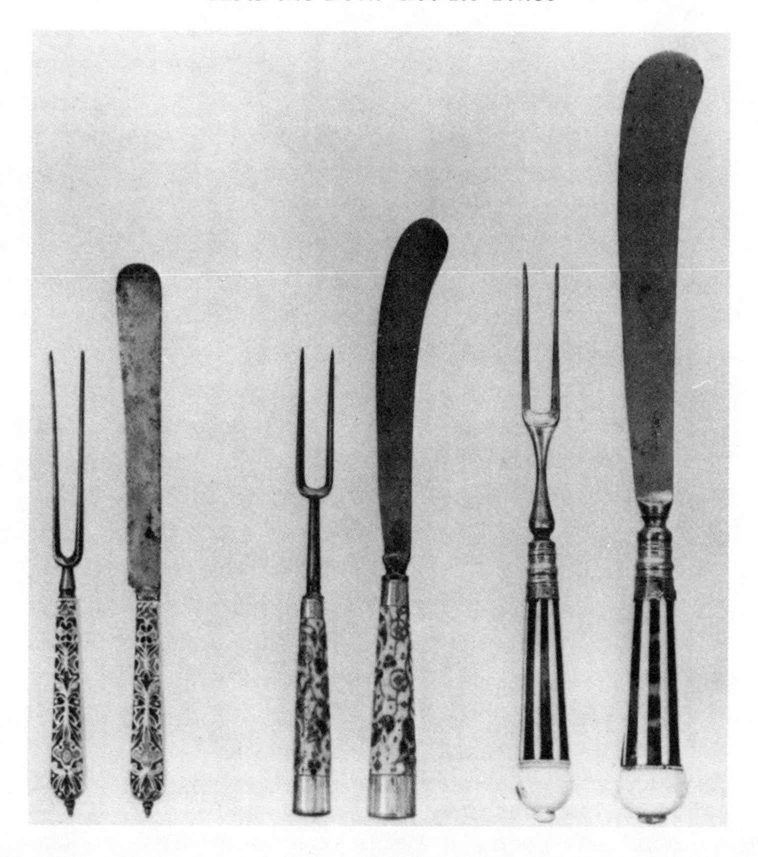

Early two-tined forks worked well for holding meat being cut but were not useful for scooping up peas and other loose food. The bulbous tip of the knife blade evolved to provide an efficient means of conveying food to the mouth, with the curve of the blade reducing the amount of wrist contortion needed to use the utensil thus. These English sets date (left to right) from approximately 1670, 1690, and 1740.

evolved into the scooper and shoveler of food, and the knife was to be reserved for cutting. The blunt-nosed straight-bladed knife, which was often more efficient as a spreading than a cutting utensil, remained in fashion throughout the nineteenth century. However, unless the cutting edge of the blade extended some distance below the line of the handle around which the fingers curled, only the tip of the blade was fully practical for cutting and slicing. This shortcoming caused the knife's bottom edge to evolve into the convex shape of most familiar table knives of today. The top edge serves no

With the introduction of three- and four-tined forks, the latter sometimes called "split spoons," it was no longer necessary or fashionable to use the knife as a food scoop, and so its bulbous curved blade reverted to more easily manufactured shapes. However, habit and custom persisted at the dinner table, and the functionally inefficient knife was used throughout the nineteenth century by less refined diners for putting food in the mouth. Left to right, these sets date from about 1805, 1835, and 1880.

purpose other than stiffening the blade against bending, and since this has not been found to be wanting, there has been essentially no change in the shape of that edge of the knife for two centuries.

Whereas the shapes of table knives have evolved to remove their existing failings and shortcomings, kitchen knives have changed little over the centuries. Their blades have remained pointed, the shape into which they naturally evolved by successive correction of faults from flint shards. The inadequacy of the common table knife to be all things to all people is emphasized when we eat a food like steak. Since the table knife is generally not sharp-pointed enough to work its way in tight curves around pieces of gristle and bone, we are brought special implements that are more suited to the task at hand. Cutting up a steak is very much like kitchen work, and so the steak knife has evolved back from the table knife to look like a kitchen knife.

The modern table knife and fork have evolved through a kind of symbiotic relationship, but the general form of the spoon has developed more or less independently. The spoon is sometimes claimed to be the first eating utensil, since solid food could easily be eaten with the bare fingers and the knife is thought to have had its beginnings as a tool or weapon rather than as an eating utensil per se. It is reasonable to assume that the cupped hand was the first spoon, but we all know how inefficient it can be. Empty clam, oyster, or mussel shells can be imagined to have been spoons, with distinct advantages over the cupped hand or hands. Shells could hold liquid longer than cramping hands, and they enabled the latter to be kept clean and dry. But shells have their own shortcomings. In particular, it is not easy to fill a shell from a bowl of liquid without getting the fingers wet, and so a handle would naturally have been added. Spoons formed out of wood could incorporate a handle integrally, and the very word "spoon" comes from the Anglo-Saxon "spon," which designated a splinter or chip of wood. With the introduction of metal casting to make spoons, the shape of bowls was not limited to those naturally occurring in nature and thus could evolve freely in response to real or perceived shortcomings, and to fashion. But even having been shaped, from the fourteenth century to the twentieth, successively round, triangular (with the handle at the apex, sometimes said to be fig-shaped), elliptical, elongated triangular (with the handle at the base), ovoid, and elliptical, the bowl of the spoon has never been far from the shape of a shell.

The use of the knife, fork, and spoon in late-seventeenth- and early-eighteenth-century Europe has influenced some persistent differences in their use by Europeans and Americans today. The introduction of the fork produced an asymmetry in tableware, and the question of which implement a diner's right and left hand held could no longer be considered moot. With identical knives in each hand, the diner was able to cut and carry food to the mouth with either knife, but, whether by custom or natural inclination, right-handedness may be assumed always to have prevailed, and so the knife in the right hand not only performed the cutting, which took much more dexterity than merely holding the meat steady on the plate, but also speared the cut-off morsel to convey it to the mouth. Because it did not need to be pointed, the left-hand knife was sometimes blunt-ended and used as a spatula to scoop up looser food or slices of meat. When the fork gained currency, it displaced the noncutting and relatively passive knife in the left hand, and in time the function of the knife in the right hand changed. With its point blunted, it was used only as a cutter and shoveler, and the fork held meat that was being cut and speared it for lifting to the mouth, a relatively easy motion with the left hand, even for a right-handed person.

By the eighteenth century, the European style of using utensils had become somewhat standardized, with the knife in the right hand cutting off food and sometimes also pushing pieces of it onto the fork, which conveyed it to the mouth. Since the first forks were straight-tined, there was no front or back to them, but shortcomings of this ambiguous design soon became evident. Whether food was skewered on or placed across the tines of the fork, the fork had to be brought to a near-horizontal position to enter the mouth with the least chance of its tines' piercing the roof of the mouth or the food's falling off. With slightly curved tines, and with food placed on their convex side, the fork handle did not have to be lifted so high to convey the food quickly and safely to the mouth. Furthermore, the arching tines enabled the fork to pierce a piece of meat squarely, yet curved out of the way so that diners could see clearly what they were cutting. By the middle of the eighteenth century, gently curving tines were standard on English forks, thus giving them distinct fronts and backs.

But the fork was a rare item in colonial America. According to one description of everyday life in the Massachusetts Bay Colony, the first and only fork in the earliest days, carefully preserved in its case, had been brought over in 1630 by Governor Winthrop. In seventeenth-century America, "knives, spoons, and fingers, with plenty of napery, met the demands of table manners." As the eighteenth century dawned, there were still few forks. Furthermore, since knives imported from England had ceased to come with pointed tips, they could not be employed to spear food and convey it to the mouth.

How the present American use of the knife and fork evolved does not seem to be known with certainty, but it has been the subject of much speculation. Without forks, the more refined colonists can be assumed to have handled a knife and spoon at the dinner table. Indeed, using an older, pointed knife and spoon, a "spike and spon," to keep the fingers from touching food may have given us the phrase "spic and span" to connote a high standard of cleanliness. How the blunted spike and spon influenced today's knife and fork has been suggested by the archaeologist James Deetz, who has written of Early American life in his evocative *In Small Things Forgotten.* (The phrase is taken from colonial probate records, where it referred to the completion of an accounting of an estate's items by grouping together the small and trivial things whose individually intrinsic value did not warrant a separate accounting. Forks themselves would never have been lumped with "small things forgotten," but still the way knives, forks, or spoons were actually used seems not to have been recorded.)

According to Deetz, in the absence of forks some colonists took to holding the spoon in the left hand, bowl down, and pressing a piece of meat against the plate so that they could cut off a bite with the knife in the right hand. Then the knife was laid down and the spoon transferred from the left to the generally preferred hand, being turned over in the process, to scoop up the morsel and transfer it to the mouth (the rounded back of a spoon being ill suited to pile food upon). When the fork did become available in America, its use replaced that of the spoon, and so the customary way of eating with a knife and spoon became the way to eat with a knife and fork. In particular, after having used the knife to cut, the diner transferred the fork from the left to the right hand, turning it over in the process, to scoop up the food for the mouth, for the spoonlike scooping action dictated that the fork have the tines curving upward. This theory is supported by the fact that when the four-tined fork first appeared in America it was sometimes called a "split spoon." The action of passing the fork back and forth between hands, a practice that Emily Post termed "zigzagging" and contrasted to the European "expert way of eating," persists to this day as the American style.

In America as elsewhere, however, well into the nineteenth century table manners and tableware remained far from uniform. Though "etiquette manuals appeared in unprecedented numbers," as late as 1864 Eliza Leslie could still declare in her *Ladies' Guide to True Politeness and Perfect Manners* that "many persons hold silver forks awkwardly, as if not accustomed to them." Frances Trollope described among the diners on a Mississippi River steamboat in 1828 some "generals, colonels, and majors" who had "the frightful manner of feeding with their knives, till the whole blade seemed to enter the mouth." And since the feeding knife was apparently blunt-tipped, the diners had to clean their teeth with pocket knives afterward. Just a generation later, the experiences of Mrs. Trollope's son, Anthony, were quite different. Dining in a Lexington, Kentucky, hotel in 1861, he observed not officers but "very dirty" teamsters who nevertheless impressed him by being "less clumsy with their knives and forks . . . than . . . Englishmen of the same rank."

On an American tour in 1842, Charles Dickens noted that fellow passengers on a Pennsylvania canal boat "thrust the broad-bladed knives and the two-pronged forks further down their throats than I ever saw the same weapons go before, except in the hands of a skilled juggler." The growing use of the fork displaced the knife from the mouth, but the new fashion was not without its dissenters, who likened eating peas with a fork to "eating soup with a knitting needle." With its multiplying tines and uses, however, the fork was to become the utensil of choice, and by the end of the nineteenth century a refined person could eat "everything with it except afternoon tea." It was just such a menu of applications for a single utensil that led to specialized descendants like fish and pastry forks, as we shall see later in this book.

European and American styles of eating with knife and fork are not the only ways civilized human beings have solved the design problem of getting food from the table to the mouth. Indeed, as Jacob Bronowski pointed out, "A knife and fork are not merely utensils for eating. They are utensils for eating in a society in which eating is done with a knife and fork. And that is a special kind of society." To this day, some Eskimos, Africans, Arabs, and Indians eat with their fingers, observing ages-old customs of washing before and after the meal. But even Westerners sometimes eat with their fingers. The American hamburger and hot dog are consumed without the aid of utensils, with the bun keeping the fingers from becoming greasy. Tacos may be less easy to eat, but the shell—reminiscent of the first food containers—keeps the greasier food from soiling the fingers, at least in principle. Such foods demonstrate alternative technological ways of achieving the same cultural objective.

In the Far East, chopsticks developed about five thousand years ago as extensions of the fingers. According to one theory of their origin, food was cooked in large pots, which held the heat long after everything was ready to be eaten. Hungry people burned their fingers reaching into the pot early to pull out the choicest-looking morsels, and so they sought alternatives. Grasping the morsels with a pair of sticks protected the fingers, or so one tradition has it. Another version credits Confucius with advising against the use of knives at the table, for they would remind the diners of the kitchen and the slaughterhouse, places the "honorable and upright man keeps well away from." Thus Chinese food has traditionally been prepared in bite-sized pieces or cooked to sufficient tenderness so that pieces could be torn apart with the chopsticks alone.

Just as Western eating utensils evolved in response to real and perceived shortcomings, so a characteristic form of modern chopsticks, rounded at the food end and squarish at the end that fits in the hand, no doubt evolved over the course of time because rounded sticks taken from nature left something to be desired. Whereas any available twigs may have served well the function of grasping food from a common pot, they would not have seemed so appropriate for dining in more formal settings. The obvious way to imitate twigs to make better chopsticks would be to form wood into straight, round rods of the desired size. But such an apparent improvement might also have highlighted shortcomings overlooked in the cruder implements. Finely shaped chopsticks that were of the same diameter at both the food and the finger ends might prove to be too thick to tear apart certain foods easily, or too thin to be comfortable during a longish meal. Thus, it would have been an obvious further improvement to make the sticks tapered, with the different ends becoming fixed at compromise sizes that made them function better for both food and hand. Whether uniform or tapered, however, round chopsticks would tend to twist in the fingers and roll off the table, and so squaring one end eliminated two annoyances in what is certainly a brilliant design.

Putting implements as common as knife and fork and chopsticks into an evolutionary perspective, tentative as it necessarily must be, gives a new slant to the concept of their design, for they do not spring fully formed from the mind of some maker but, rather, become shaped and reshaped through the (principally negative) experiences of their users within the social, cultural, and technological contexts in which they are embedded. The formal evolution of artifacts in turn has profound influences on how we use them.

Imagining how the form of things as seemingly simple as eating utensils might have evolved demonstrates the inadequacy of a "form follows function" argument to serve as a guiding principle for understanding how artifacts have come to look the way they do. Reflecting on how the form of the knife and fork has developed, let alone how vastly divergent are the ways in which Eastern and Western cultures have solved the identical design problem of conveying food to mouth, really demolishes any overly deterministic argument, for clearly there is no unique solution to the elementary problem of eating.

What form does follow is the real and perceived failure of things as they are used to do what they are supposed to do. Clever people in the past, whom we today might call inventors, designers, or engineers, observed the failure of existing things to function as well as might be imagined. By focusing on the shortcomings of things, innovators altered those items to remove the imperfections, thus producing new, improved objects. Different innovators in different places, starting with rudimentary solutions to the same basic problem, focused on different faults at different times, and so we have inherited culture-specific artifacts that are daily reminders that even so primitive a function as eating imposes no single form on the implements used to effect it.

The evolution of eating utensils provides a strong paradigm for the evolution of artifacts generally. There are clearly technical components to the story, for even the kind of wood in chopsticks or the kind of metal in knives and forks will have a serious impact on the way the utensils can be formed and can carry out their functions. Technological advances can have far-reaching implications for the manner of manufacture and use of utensils, as the introduction of stainless steel did for tableware, which in turn can affect their price and availability across broad economic classes of people. But the stories associated with knives, forks, and spoons also illustrate well how interrelated are technology and culture generally. The form, nature, and use of all artifacts are as influenced by politics, manners, and personal preferences as by that nebulous entity, technology. And the evolution of the artifacts in turn has profound influences on manners and social intercourse.

But how do technology and culture interact to shape the world beyond the dinner table? Are there general principles whereby all sorts of things, familiar and unfamiliar, evolve into their shapes and sizes and systems? If not in tableware, does form follow function in the genesis and development of our more high-tech designs, or is the alliterative phrase just an alluring consonance that lulls the mind to sleep? Is the proliferation of made things, such as the seemingly endless line of serving pieces that complement a table service, merely a capitalist trick to sell consumers what they do not need? Or do artifacts multiply and diversify in an evolutionary way as naturally as do living organisms, each having its purpose in some wider scheme of things? Is it true that necessity is the mother of invention, or is that just an old wives' tale? These are among the questions that have prompted this book. In order to begin to answer them, it will be helpful first to look beyond a place setting of examples to rules, and then to illustrate them by an omnivorous selection of further examples. Thus is the design problem of this literary artifact.

2

Form Follows Failure

The evolution of the modern knife and fork from flint and stick, and the evolution of the spoon from the cupped hands and shells of eons ago, seem thoroughly reasonable stories. But they are more than stories, constructed after the fact by imaginative social scientists; the way our common tableware has developed to its present form is but a single example of a fundamental principle by which all made things come to look and function the way they do. That principle revolves about our perception of how existing things fail to do what we expect them to do as well and conveniently and economically as we think they should or wish they would. In short, they leave something to be desired.

But whereas the shortcomings of an existing thing may be expressed in terms of a *need* for improvement, it is really *want* rather than need that drives the process of technological evolution. Thus we may need air and water, but generally we do not require air conditioning or ice water in any fundamental way. We may find food indispensable, but it is not necessary to eat it with a fork. Luxury, rather than necessity, is the mother of invention. Every artifact is somewhat wanting in its function, and this is what drives its evolution.

Here, then, is the central idea: the form of made things is always subject to change in response to their real or perceived shortcomings, their failures to function properly. This principle governs all invention, innovation, and ingenuity; it is what drives all inventors, innovators, and engineers. And there follows a corollary: Since nothing is perfect, and, indeed, since even our ideas of perfection are not static, everything is subject to change over time. There can be no such thing as a "perfected" artifact; the future perfect can only be a tense, not a thing.

If this hypothesis is universally valid and can explain the evolution of all made things, it must apply to any artifact of which we can think. It must explain the evolution of the zipper no less than the pin; the aluminum can no less than the hamburger package; the suspension bridge no less than Scotch tape. The hypothesis fleshed out must also have the potential for explaining why some of our most everyday things continue to look the way they do in spite of all their obvious shortcomings. It must explain why some things change for the worse, and why those things aren't made in the good old way. Some background from the writings of inventors and designers and those who think about invention and design can set the stage for the case studies that will test the hypothesis.

The large number of things that have been devised and made by humans throughout the ages has been estimated in some recently published books on the design and evolution of artifacts. Donald Norman, in *The Design of Everyday Things,* describes sitting at his desk and seeing about him a host of specialized objects, including various writing devices (pencils, ballpoint pens, fountain pens, felt-tip markers, highlighters, etc.), desk accessories (paper clips, tape, scissors, pads of paper, books, bookmarks, etc.), fasteners (buttons, snaps, zippers, laces, etc.), etc. In fact, Norman counted over one hundred items before he tired of the task. He suggests that there are perhaps twenty thousand everyday things that we might encounter in our lives, and he quotes the psychologist Irving Biederman as estimating that there are probably "30,000 readily discriminable objects for the adult." The number was arrived at by counting the concrete nouns in a dictionary.

George Basalla, in *The Evolution of Technology,* suggests the great "diversity of things made by human hands" over the past two hundred years by pointing out that five million patents have been issued in America alone. (Not every new thing is patented, of course, and we can get some idea of the enormity of our re-arrangement and processing of things by noting that over ten million new chemical substances were registered in the American Chemical Society's computer data base between 1957 and 1990.) Basalla also notes that, in support of Darwin's evolutionary theory, biologists have identified and named over one and a half million species of flora and fauna, and thus he concludes that, if each American patent were "counted as the equivalent of an organic species, then the technological can be said to have a diversity three times greater than the organic." He then introduces the fundamental questions of his study:

> The variety of made things is every bit as astonishing as that of living things. Consider the range that extends from stone tools to microchips, from waterwheels to spacecraft, from thumbtacks to skyscrapers. In 1867 Karl Marx was surprised to learn . . . that five hundred different kinds of hammers were produced in Birmingham, England, each one adapted to a specific function in industry or the crafts. What forces led to the proliferation of so many variations of this ancient and common tool? Or more generally, why are there so many different kinds of things?

Basalla dismisses the "traditional wisdom" that attributes technological diversity to necessity and utility, and looks for other explanations, "especially ones that can incorporate the most general assumptions about the meaning and goals of life." He finds that his search "can be facilitated by applying the theory of organic evolution to the technological world," but he acknowledges that the "evolutionary metaphor must be approached with caution," because fundamental differences exist between the made world and the natural world. In particular, Basalla admits that, whereas natural things arise out of random natural processes, made things come out of purposeful human activity. Such activity, manifested in psychological, economic, and other social and cultural factors, is what creates the milieu in which novelty appears among continuously evolving artifacts.

Adrian Forty has also reflected on the multitude of made objects. In *Objects of Desire,* he notes that historians have generally accounted for the diversification of designs in one of two ways. The first explanation, albeit a rather circular one, is that there is an ongoing evolution of new needs created by the development of new designs, such as machinery and appliances that are increasingly complicated and compact. The new designs require new tools for assembly and disassembly, and these new tools in turn enable still further new designs to be realized. The second explanation for the diversity of artifacts is "the desire of designers to express their ingenuity and artistic talent." Both theories were used by Siegfried Giedion in *Mechanization Takes Command,* but neither theory, convincing though it may be in explaining particular cases of diversity, covers all cases, as Forty admits.

In mid-nineteenth-century America, for example, there developed a new piece of furniture, the adjustable chair. Giedion's explanation for the proliferation of designs for such a chair was that it was prompted by the posture of the times, which was based on relaxation, "found in a free, unposed attitude that can be called neither sitting nor lying." He argued that the development of the new patent furniture was thus in response to a new need, which happily coincided with a concentration of creativity among ingenious designers. But Forty rejects Giedion's reasoning as overly dependent on coincidence, arguing that "it is most unlikely that after several millennia mankind should suddenly have discovered a new way of sitting in the nineteenth century," when "designers were no more inventive and ingenious than people at other times."

Forty dismisses the "functionalist" theory as inadequate to explain the diversity of a less adjustable but more recent example: "Could Montgomery Ward's 131 different designs of pocket knife be said to be the result of the discovery of new ways of cutting?" And he does not allow that nineteenth-century designers, no matter how ingenious, had the power or autonomy to influence "how many or what type of articles should be made," although he does agree that designers could determine the form of individual articles. Forty's own arguments for the multiplication of things like adjustable chairs "place the products of design in a direct relationship to the ideas of the society in which they are made." In particular, he identifies the capitalists as the proliferators of diversity: "The evidence is that manufacturers themselves made distinctions between designs on the basis of different markets." Thus there exists a dictionary situation for everyone: designers design, manufacturers manufacture, and diverse consumers consume diversity. This is or is not a nefarious arrangement, depending upon one's ideology.

Whether or not the world should have diversity, it does, and the question remains as to how individual designs come to be distinguished from related designs. Even if manufacturers are the primary driving force for diversity, what underlying idea governs how a particular product looks? Certainly it was more than economic considerations alone that distinguished one from the other among those 131 knives in the Montgomery Ward's catalogue, one from the other among those five hundred specialized hammers made in Birmingham. Certainly there were distinctions, but what forces created them?

Neither Norman, Basalla, nor Forty has much to say about a relationship between form and function. The words do not appear in any of their indexes, and we can confidently assume that these authors do not subscribe to the formula "form follows function," which Forty calls an "aphorism." Nor does David Pye, who has written very cogently about design. Pye's books are especially rewarding reading because he lets the reader see how he thinks. He does not just give us the polished fruits of his thought; he also gives us the pits and seeds and cores, so that we may observe what is at the heart of his thinking through a design problem. Not only does he dismiss "form follows function" as "doctrine," he also ridicules the dictionary definition that function is "the activity proper to a thing."

According to Pye, "function is a fantasy," and he italicizes his further assertion that "the form of designed things is decided by choice or else by chance; but it is never actually entailed by anything whatever." He ridicules the idea that something "looks like that because it has got to be like that," and equates "purely functional" with terms that to him are pejoratives, such as "cheap" and "streamlined." He elaborates on his disdain for the idea that "form follows function":

> The concept of function in design, and even the doctrine of functionalism, might be worth a little attention if things ever worked. It is, however, obvious that they do not. Indeed, I have sometimes wondered whether our unconscious motive for doing so much useless work is to show that if we cannot make things work properly we can at least make them presentable. Nothing we design or make ever really works. We can always say what it ought to do, but that it never does. The aircraft falls out of the sky or rams the earth full tilt and kills the people. It has to be tended like a new born babe. It drinks like a fish. Its life is measured in hours. Our dinner table ought to be variable in size and height, removable altogether, impervious to scratches, self-cleaning, and having no legs. . . . Never do we achieve a satisfactory performance. . . . Every thing we design

> and make is an improvisation, a lash-up, something inept and provisional.

Pye is engaging in hyperbole, of course, but all hyperbole has its roots in truth. What is at the root of Pye's ranting is that nothing is perfect: If a malfunction occurs in one out of a million airline flights, then the aircraft is not perfected in the strictest sense of the word. Only tending to airplanes as if they were newborn babes keeps them well enough maintained to hold accident rates down. The truly perfected airplane would not need maintenance, would fly on little fuel, and would last for centuries, if not longer. And what is wrong with the dinner table? Well, we do have to insert and remove leaves to accommodate our variable-sized dinner parties. We have to situate telephone books to bring the latest generation up to table height. The table does just sit there when we are not using it. Its finish gets scratched, and it gets dirty. And it has legs that restrict our movement up to and away from it. In short, the table, like all designed objects, leaves room for improvement.

In fact, it is just this ubiquitous imperfection that Pye so exaggerates that is the single common feature of all made objects. And it is exactly this feature that drives the evolution of things, for the coincidence of a perceived problem with an imagined solution enables a design change. But such a scenario for the evolution of artifacts should give us ever better designs, yet it does not seem to do so. A resolution of the paradox lies in Pye's observation that design requirements are always in conflict and hence "cannot be reconciled":

> All designs for devices are in some degree failures, either because they flout one or another of the requirements or because they are compromises, and compromise implies a degree of failure. . . .
>
> It follows that all designs for use are arbitrary. The designer or his client has to choose in what degree and where there shall be failure. Thus the shape of all things is the product of arbitrary choice. If you vary the terms of your compromise—say, more speed, more heat, less safety, more discomfort, lower first cost—then you vary the shape of the thing designed. It is quite impossible for any design to be "the logical outcome of the

requirements" simply because, the requirements being in conflict, their logical outcome is an impossibility.

Thus the common dinner table that Pye had described earlier is a failure because it cannot meet simultaneously all the competing terms of seating two and twelve, seating small children and large adults, possessing an aesthetically pleasing finish that does not scratch or soil, and having legs that hold it up without getting in the way. We can find fault with any common object if we look hard enough at it. But that is not Pye's goal, nor is it this book's intention. Rather, the objective here is to celebrate the clever and everyday things of an imperfect world as triumphs in the face of design adversity. We will come to understand why we can speak of "perfected" designs in such an environment, and why one thing follows from another through successive changes, all *intended* to be for the better.

Few writers have been more explicit about the role of failure in the evolution of artifacts than the architect Christopher Alexander in his *Notes on the Synthesis of Form.* He makes it abundantly clear that we must look to failure if we ever hope to declare success, and he illustrates the principle with the example of how a metal face can be declared "perfectly" smooth and level. We can ink the face of a standard block that is known to be level and rub it against the face being machined:

> If our metal face is not quite level, ink marks appear on it at those points which are higher than the rest. We grind these high spots, and try to fit it against the block again. The face is level when it fits the block perfectly, so that there are no high spots which stand out any more.

A dentist fitting a crown employs a similar technique. Although he does not seek a totally level surface, the dentist does want the new tooth surface to conform to its mate. This is done by having the patient grind something like carbon paper between the teeth to mark those high spots where the crown fails to fit. It is clear from Alexander's paradigm for realizing the design of an artifact, which to him consists of fitting form to context, that we are able to declare success only when we can find no more points that fail to conform to the standard against which we judge. In general, a successful design, which Alexander terms a good fit between form and context, can be declared only when we can detect no more differences. It is "departures from the norm which stand out in our minds, rather than the norm itself. Their wrongness is somehow more immediate than the rightness."

Alexander also gives a more everyday example, one that does not require a machine shop or a dentist's office to re-enact. All we need is a box of buttons that have collected over the years:

> Suppose we are given a button to match, from among a box of assorted buttons. How do we proceed? We examine the buttons in the box, one at a time; but we do not look directly for a button which fits the first. What we do, actually, is to scan the buttons, rejecting each one in which we notice some discrepancy (this one is larger, this one darker, this one has too many holes, and so on), until we come to one where we can see no differences. Then we say that we have found a matching one.

This is essentially how a word processor's spelling-checker program can work. It takes each word in a document in turn and compares it with the words in its dictionary. The logic or software of the program can find a matching word, if any exists, by successively eliminating those that do not match. Words of different length from the one being checked can be eliminated first because they obviously can't fit letter for letter. Then the remaining words in the dictionary that do not have the same first letter as the word being checked can be eliminated. Of those words remaining, those that do not have the same second letter can be eliminated, and so forth until the last letter in the word being checked is reached. If there remains a word in the dictionary that produces no misfits, then the word whose spelling is being checked can be declared correctly spelled. If all the words in the dictionary are found to be misfits, then the word being checked can be declared misspelled. The success of the program depends fundamentally on the concept of failure. (The logic has several shortcomings, of course, which if not dealt with separately will not catch certain misspelled words and will declare some correctly spelled words misspelled. For example, it will not catch such misspellings as "their" for "there," because both are valid words in the dictionary.)

Alexander generalizes from his examples to recommend that "we should always expect to see [design] as a negative process of neutralizing the incongruities, or irritants, or forces, which cause misfit" between form and context. This is also how artifacts change over time and evolve with use. The manipulation of two pointed knives to eat a piece of meat might frequently have irritated medieval diners as the meat rotated about the stationary knife. Diners who chose not to touch the meat with their fingers would generally have been able to neutralize the irritant by pressing the noncutting knife flatter onto the meat, thus altering its use. And in time this might change the form of the knife blade to give it a better bearing surface. Knife makers are also diners, of course, and a particularly reflective or imaginative one might have come up with a more radical way of neutralizing the irritant—developing an entirely different eating implement, one with two prongs to stab the meat in order to prevent it from rotating while being sliced.

"Misfit provides an incentive to change; good fit provides none," declares Alexander, and even if we ourselves do not have the material, tools, or ability to work up a new artifact to remove an irritant in one we use, we might at least call the irritant to the attention of someone who can. That someone who can effect changes can be a craftsperson, for whom Alexander uses the masculine to include the feminine, and whom he describes as "an agent simply" through whom artifacts can evolve in an almost organic way:

> Even the most aimless changes will eventually lead to well-fitting forms, because of the tendency to equilibrium inherent in the organization of the process. All the agent need do is recognize failures when they occur, and to react to them. And this even the simplest man can do. For although only few men have sufficient integrative ability to invent form of any clarity, we are all able to criticize existing forms. It is especially important to understand that the agent in such a process needs no creative strength. He does not need to be able to improve the form, only to make some sort of change when he notices a failure. The changes may not be always for the better; but it is not necessary that they should be, since the operation of the process allows only the improvements to persist.

This evolutionary process has worked throughout civilization and continues to work today even as craftsmen have become scientifically savvy engineers, and even as artifacts have grown to the complexity of nuclear-power plants, space shuttles, and computers. However, unlike Alexander's agent, who does not necessarily have to make changes for the better, when the modern designer or inventor makes a change in an artifact, he or she must definitely think it is for the better in some sense. Nevertheless, actual incidents as well as mere perceptions of misfit and failure do continue to drive the evolution of artifacts, and we can expect that they always will. And it need not be only the likes of engineers, politicians, and entrepreneurs who have a hand in shaping the world and its things, for we are all specialists in at least a small corner of the world of things. We are all perfectly capable of seeing what fails to live up to the promise of its designers, makers, sellers, or licensers. Such ideas should be as evident to users of artifacts today as they were to the governed in the days of the Athenian statesman Pericles, who observed that "although only a few may originate a policy, we are all able to judge it."

Understanding how and why our physical surroundings come to look and work the way they do provides considerable insight into the nature of technological change and the workings of even the most complex of modern technology. Basalla takes the artifact as "the fundamental unit for the study of technology," and argues convincingly that "continuity prevails throughout the made world." Thus the cover illustration for *The Evolution of Technology* depicts "the evolutionary history of the hammer, from the first crudely shaped pounding stone to James Nasmyth's gigantic steam hammer," with which steel forgings of unprecedented size could be made at the culmination of the Industrial Revolution. Basalla asserts that the existence of such continuity in all things "implies that novel artifacts can only arise from antecedent artifacts—that new kinds of made things are never pure creations of theory, ingenuity, or fancy." If this be so, then the history of artifacts and technology becomes more than a cultural adjunct to the business of engineering and invention. It becomes a means of understanding the elusive creative process itself, whereby the intellectual capital of nations is generated.

The same purposeful human activity that has shaped such common objects as the knife, fork, and spoon shapes all objects of technology, "from stone tools to microchips," and also accounts for the diversity of things, from the hundreds of hammers made in nineteenth-century Birmingham to the multitude of specialized pieces of silverware that came to constitute a table service. The distinctly human activities of invention, design, and development are themselves not so distinct as the separate words for them imply, and in their use of failure these endeavors do in fact form a continuum of activity that determines the shapes and forms of every made object.

Whereas shape and form are the fundamental subjects of this book, the aesthetic qualities of things are not among its primary concerns. Aesthetic considerations may certainly influence, and in some cases even dominate, the process whereby a designed object comes finally to look the way it does, but they are seldom the first causes of shape and form, with jewelry and objets d'art being notable exceptions. Utilitarian objects can be streamlined and in other ways made more pleasing to the eye, but such changes are more often than not cosmetic to a mature or aging artifact. Tableware, for example, has clearly evolved for useful purposes, and, no matter what pattern of silver we may have set before us on a table, we do not confuse knife, fork, and spoon. But when aesthetic considerations dominate the design of a new silverware pattern, the individual implements, no matter how striking and well balanced they may look on the table, can often leave much to be desired in their feel and use in the hand. Chess pieces constitute another example of a set of objects that have long-established and fixed requirements. There is no leeway as to how many pawns or rooks a set must have, and there is no getting around the fact that the pieces must be distinguishable one from another and must be provided in two equal but easily separated groups of sixteen. To design or "redesign" a chess set may involve some minor considerations of weight and balance in the pieces, but more often than not it is taken as a problem in aesthetics. And in the name of aesthetics many a chess set has been made more modern- or abstract-looking, if not merely different-looking, at the expense of chess players' ability to tell the queen from the king or the knight from the bishop. Such design games are of little concern in this book.

We shall, however, be concerned with what is variously called "product design" or "industrial design." Though this activity often appears to have aesthetics as its principal consideration, the best of industrial design does not have so narrow a focus. Rather, the complete industrial designer seeks to make objects easier to assemble, disassemble, maintain, and use, as well as to look at. And the best of industrial designers will have the ability to see into the future of a product so that what might have been a damning shortcoming of an otherwise wonderful-looking and beautifully functioning artifact will be nipped in the bud. Considerations that go variously under the name "human-factors engineering" or, especially in England, "ergonomics" are closely related to those of industrial design, but the human-factors engineer is especially concerned with how anything from the simplest kitchen gadget to the most advanced technological system will behave at the hands of its intended, and perhaps unintended, users.

The childproof bottle for prescription medicine is something that many people, especially older folks with arthritis, would agree could benefit from some industrial redesign, but most would also no doubt concur that the focus should be on the human-engineering aspects of getting the top off before the container gets an aesthetic treatment, although that would also be welcome. The ideal prescription-medicine container would be human-engineered to perfection and yet attractive enough to displace a bowl of fruit on the kitchen table. Such pretty things may not be designed in this book, but the intention is to go at least some way toward developing an understanding of why such things do not exist among the myriad ones that do. Just as there are many ways in which an artifact can fail, so there are many paths that a corrective form can follow.

3

Inventors as Critics

If the shortcomings of things are what drive their evolution, then inventors must be among technology's severest critics. They are, and it is the inventor's unique ability not only to realize what is wrong with existing artifacts but also to see how such wrongs may be righted in order to provide increasingly more sophisticated gadgets and devices. These contentions are not just the wishful thinking of a theorist seeking order in the made world; they are grounded in the words of reflective inventors themselves, who come from all walks of life.

Jacob Rabinovich was the son of a Russian shoe manufacturer who moved his family to Siberia at the outbreak of war in 1914. After five years, when Jacob was eleven years old, the family emigrated to America, and they settled in New York City. Young Jacob was a well-rounded student in high school, belonging to both the mathematics team and the drawing team. His freehand work was admired for its accuracy by the head of his school's art department, but the teacher advised the young man to study engineering, because his drawings lacked spirit. City College of New York was the institution of opportunity and hence of choice for many a young immigrant child in the 1920s; however, Jacob was advised that the engineering profession was a difficult one to break into, especially for a Jew. So he entered City College in 1928 as a general-studies major and suddenly found himself to be a mediocre student amid strong competition.

The coming of the Great Depression meant that it would be difficult to earn a living in any line of work, and so Jacob changed his major to engineering, his first love. By 1933, when he graduated with a degree in electrical engineering, he had Americanized his name to "Jacob Rabinow." He stayed on at City College for an extra year to earn the equivalent of a master's degree, but even then jobs were hard to come by, and he spent some years working in radio factories, much of the time doing assembly work. He took a Civil Service examination in 1935, receiving high marks on both the electrical- and mechanical-engineering parts, but he did not land a government job until 1938, as an engineer in the National Bureau of Standards, where his first duties consisted of calibrating instruments used to measure the rate of water flow in streams and rivers.

Rabinow found his first assignment not unpleasant, and the routineness of the work gave him plenty of time to think. The equipment he was using was old and worn, with many shortcomings, and he soon began to see various ways to improve its operation and accuracy. He approached his boss, who had no objection to Rabinow's designing and building new equipment, as long as it was done after hours, on his own time. He soon came up with some obviously improved calibrating equipment and also began to show his talents in other areas, and so he was given increasing responsibility and independence—and he flourished, eventually owning his own company for a while. All in all, Rabinow holds 225 patents for devices ranging from self-regulators for watches and clocks to the automatic letter-sorting machines used by the Postal Service.

Over the course of his career Rabinow, in the characteristic fashion of engineers and inventors, wrote relatively little for general publication. But in retirement he published his first full-length book, *Inventing for Fun and Profit,* which in spite of its title provides a highly original and revealing insight into the inventor's mind. The origins of many of Rabinow's inventions are described, and those origins lie typically in finding fault with existing things. Thus he relates such stories as how the difficulty in adjusting a watch he received as a present led him to invent a self-adjusting watch, or how his arguing with a fellow music lover over whether or not the sound issuing from conventional phonographs was distorted (because of the way the arm constrained the needle to move in the record groove) led to his development of a new needle arm suspension system. Problems brought to him by friends proved to be an especially fertile source of ideas for new projects. With a prolific inventor like Rabinow, there appears to be little separation between home, social, and professional life, as testified to by the location of his home workshop just off his living room.

Sometimes Rabinow gets quite explicit about the nature of his type: "Inventors are people who not only curse, but who also start to think of what can be done to eliminate the bother." He repeated this view when asked why he invents. Rabinow responded, "When I see something that I don't like, I try to invent a way around it. My job is simply to design gadgets that I like." Of course, gadgets that he likes will not have the faults of those he found wanting. Many inventors echo Rabinow's identification of shortcomings as the driving force for change. Lawrence Kamm, who dedicated his book, *Successful Engineering,* to Jacob Rabinow ("my boss, teacher, close friend, and severest critic"), advises the young design engineer to "continually study the designs around you. Why were they designed as they are? What is wrong with them? How would you improve them?"

Inventors at Work, a collection of interviews with sixteen notable American inventors, provides a sampling of the variety of educational backgrounds, ranging from terminal high-school degrees to doctorates, from which the breed comes. For every notable inventor who had to go to work instead of college, there is one who was able to attend an Ivy League school. What seems more common than any educational pattern is the entrepreneurial drive, whether as an independent individual explicitly trying to turn inventions into vastly successful products or as a member of a large corporate structure pushing for innovation by working within the system.

For every scrappy immigrant inventor like Jacob Rabinow, there is one born with a silver spoon in his mouth. Paul MacCready is the creator of the Gossamer Condor, which in 1977 established the possibility of human-powered flight by completing a mile-long figure-eight course over the San Joaquin Valley. Although he freely admits that the £50,000 prize money put up by the British industrialist Henry Kremer in 1959 was a strong motivating factor for the research-and-development effort, MacCready was also attracted to the challenge because he had built model airplanes from early adolescence and by age seventeen was identified by the editors of *Model Airplane News* as "by far the most versatile model flyer," one who was always interested in discovering more efficient ways to handle old principles. He later took up the hobby of flying sailplanes, becoming national soaring champion three times.

After graduating from Yale, MacCready earned a doctorate in aeronautical engineering from the California Institute of Technology. Among his many achievements, he has been named engineer of the century by the American Society of Mechanical Engineers. But neither accolades nor prize money can keep an inveterate inventor happy. Like many successful inventors, MacCready is driven to make existing things better, and he soon developed the Gossamer Condor into the Gossamer Albatross, which crossed the English Channel under human pedal power in 1979. However, even the sagest of inventors knows that there are limitations to their talents for making good things better. When asked what challenge he would turn down, MacCready responded, "A much better bicycle. There are several avenues to follow there, and I've built some, and none of them satisfied me." Though it is implicit in this response that existing bicycle designs are imperfect, some design questions are more easily asked than definitively answered. Inventors are seldom at a loss for problems, and so they must choose which ones they will work on.

Nathaniel C. Wyeth was born in Chadds Ford, Pennsylvania, on the family homestead of the painter N. C. Wyeth. While the child's brother, Andrew, and his sisters, Henriette and Carolyn, studied art under the tutelage of their famous father, Nathaniel took clocks apart and made gadgets out of scrap metal. Originally named Newell Convers Wyeth, after his father, the young, technically inclined tinkerer soon had his first name changed to Nathaniel, after an uncle who was an engineer, so as to be less encumbered by identification with a prominent artist. He studied engineering at the University of Pennsylvania and then had a long and illustrious career with the Du Pont Corporation, culminating in 1975, when he became the first person to hold the company's highest technical position of senior engineering fellow.

Probably foremost among Wyeth's many inventions in such areas as textiles and electronics is the now ubiquitous plastic soda bottle, which he developed in the mid-1970s after extensive experiments with processing polyethylene terephthalate, more familiarly called PET. Such a bottle would have obvious advantages over the then conventional glass bottle, which was of course heavy and breakable. But the development of the PET bottle did not come easily; Wyeth recalls showing the misshapen results of an early experiment to the laboratory director, who wondered about spending so much money to get such a "terrible-looking bottle." Wyeth, who was pleased that the thing was at least hollow, persisted in his efforts, however, using, as he did in all his inventions, his "failures and the knowledge of things that wouldn't work as a springboard to new approaches." He was quite explicit about the way an idea progressed from terrible-looking things to bottles displayed proudly in supermarkets: "If I hadn't used those mistakes as stepping stones, I would never have invented anything." Whatever one may think of the plastic bottle, the thing does fulfill the objective of replacing glass bottles. That Wyeth's achievement now presents environmental problems for other inventors to solve should come as no surprise in an imperfect world of imperfect things.

Regardless of their background and motivation, all inventors appear to share the quality of being driven by the real or perceived failure of existing things or processes to work as well as they might. Fault-finding with the made world around them and disappointment with the inefficiency with which things are done appear to be common traits among inventors and engineers generally. They revel in problems—those they themselves identify in the everyday things they use, or those they work on for corporations, clients, and friends. Inventors are not satisfied with things as they are; inventors are constantly dreaming of how things might be better.

This is not to say that inventors are pessimists. On the contrary, they are supreme optimists, for they pursue innovation with the belief that they can improve the world, or at least the things of the world. Inventors do not believe in leaving well enough alone, for well enough is not good enough for them. But, also being supreme pragmatists, they realize that they must recognize limits to improvement and the trade-offs that must accompany it. Credible inventors know the limitations of the world too, including its thermodynamic laws of conservation of energy and growth of entropy. They do not seek perpetual-motion machines or fountains of youth but, rather, strive to do the best with what they have and for the best they know they can have, and they always recognize that they can never have everything.

Marvin Camras, a native Chicagoan who was educated at the Illinois Institute of Technology and spent most of his career at its affiliated research institute, holds over five hundred patents for devices in electrical communications. When once asked if he noticed whether inventors had any common traits, he responded:

> They tend to be dissatisfied with what they see around them. Maybe they're dissatisfied with something they're actually

> working on or with an everyday thing, about which they say, "Gee, this is a very poor way of doing this." At least in my case, when I see something that is clumsy or inelegant, I always wonder why it was made that way. You might say that these first ideas lead to invention. . . . A lot of things seem clumsy to me. I like to have things simplified.

Camras may be an individualistic type, as he believes inventors are generally, but his views about invention are common among his peers. Jerome Lemelson graduated with a master's degree in industrial engineering from New York University in 1951. He has designed industrial robots and automated factories and has even patented such things as cutout toys for the backs of cereal boxes. And yet, though he has more than four hundred patents to his name, Lemelson has made no attempt to become an entrepreneur, has refused to follow the familiar practice of building a company around one or more of his patents. Rather, he prefers to benefit from their royalties. His idea of how to invent also involves the criticism of existing artifacts:

> I think the way to go about it is to ask yourself these questions: Is this particular function being properly performed? Is it being performed in the best way possible? Are there any problems with it? How can I improve upon it? The patent system contains patents, most of which are simply improvements over what existed before. And that's really the name of the game: improving on what's existing today.

This idea is repeated in a "primer on inventions and patents" entitled *Money from Ideas* and published in 1950 by Popular Mechanics Press. With few pretensions to being out of the mainstream of the common dream of many an inventor, the book has its tone set in the first sentence of the first chapter: "A man once made a million dollars with a pair of scissors and a few sheets of paper." (He was a traveling salesman whose disgust with common drinking glasses in public places led him to invent the paper cup.) Whereas self-confident inventors who are also self-starters would certainly not need the assistance of such a primer, the popular image of the inventor as creative genius, national hero, and wealthy benefactor of a leisurely if not glamorous pursuit, provides ample attraction to those who may have more desire than talent to become inventors themselves. Such inventors manqués, being without their own ideas, must get them from others. In advising would-be inventors that they must accumulate an inventory of ideas, the adviser focused his attention on everyday items around the house:

> Tools! That should be worth exploring. Every household needs them. Every mechanic and workman in the country uses hand tools of some sort. Probably every workman has some pet peeve about one of his tools, something he feels should be altered or fixed. Beefing is a great American sport, and almost any mechanic should be glad to pour out a list of such troubles to an interested ear. The inventor who does not listen to others for stimulating ideas is usually a failure.

In spite of the tone and purpose of this advice, the underlying truth is universal: invention begins not so much in need as in want. The mechanic's needs are met with existing tools, and he uses his hammer, screwdriver, and wrench every working day. But his tasks vary from day to day, and his unchanging tools work better some days than others. He might have to screw some pieces of wood together to make a storage box for his workshop, or he might have to reattach a brightly finished metal panel to a machine he has repaired for a customer. (Let us assume, for the sake of argument, that this mechanic has only one conventional screwdriver, and that the wood and metal screws involved are of the conventional kind, with single slots that go across an entire diameter of the screw head.) In the one case, the screwdriver might slip out of the screw head and indent the wooden box. Though that would be unfortunate, the mechanic could no doubt live with it. In the other case, however, a slip of the screwdriver might leave a nasty scratch where the customer will not accept one. Strictly speaking, the mechanic should be able to avoid scratching the metal panel by paying close attention to how he drives the screws, carefully centering the screwdriver head firmly in the screw slot and twisting the screwdriver perfectly straight with absolutely no sideways leaning or slipping. To be extra-careful, the mechanic might even hold the fingers of one hand around the screw head so as to contain the screwdriver's head.

Such precautions would work, of course; the mechanic may once have scratched a polished panel with a screwdriver long ago, when he was less attentive, but he might never have made the same mistake again. Though we could say that the mechanic needs to be careful, he does not necessarily need a new or different screwdriver. But he would certainly take one, and inventors are always looking for opportunities to give him one. Recently, for example, screwdrivers with tungsten-carbide particles bonded to their tips have been added to the toolbox. The hard particles bite into the softer screw slots and so reduce the problem of having the screwdriver blade slip out.

Jacob Rabinow has spoken specifically about screws and screwdrivers in the context of questions he used in interviewing prospective employees. His goal was to separate theoretical scientists and engineers from practical inventors. He would observe, of the most familiar screw heads: "The slot is traditional. It's simple to make but it has several problems." In addition to the problem of the screwdriver's slipping out of the slot and damaging the work, Rabinow would mention that it is easy for people to improvise screwdrivers out of coins, nail files, and the like to remove screws that should not be removed. (This seems to have been an especially annoying habit of users of public restrooms, and so many of the screw heads in such contemplative quarters have come to be of an unusual but now familiar design that allows them to be easily installed but virtually impossible to be removed by the uninitiated.)

There are other alternatives to the conventional screw head, and Rabinow calls one, the Phillips-head screw, a "prettier design." He notes that it certainly reduces the probability of the screwdriver's slipping but, like most evolutionary designs, for every advantage it has over the traditional design it seems to have a disadvantage of its own. In the case of the Phillips-head screw, which may also have some aesthetic advantages over the common screw head, the Phillips-head screwdriver must be more closely matched to the screw head than is the case with the traditional design, and when the screwdriver gets worn with use it is much harder to sharpen than a conventional one. Rabinow displays his creativity by demonstrating how he can imagine new screw heads that eliminate some of the shortcomings of the Phillips-head screw.

He observes that there are screws made with square and hexagonal depressions in their heads, with matching screwdrivers or wrenches. He seems to prefer the design with the square depression, because "it is easy to sharpen the screwdriver and it's very positive." But of course, as Rabinow observes, "in all of these designs, a flat screwdriver of the right width can be used for unauthorized removal of a screw." Hence he asks if one can redesign the screw head so that the screws will stay put. The inventor's challenge is then: "Can one design a slot or a depression in the head of a screw that cannot be driven by a flat screwdriver of any width?" Rabinow would no doubt have hired on the spot any interviewee who responded in any way close to the master's solution:

> If you make a triangular depression with sides in the shape of three arcs, where each point of the triangle is the center of curvature of the opposite arc, you have a triangular hole that can be driven with a specially shaped screwdriver, but not by any flat screwdriver. If you insert a flat blade, the blade will pivot at each corner and slide over the opposite curved surface, hit the next corner and slide again, and so on. Such a screw should look very attractive and would be very difficult to open without the proper tool.

Rabinow admits that he does not know whether this idea is new, because he has not looked in the Patent Office files. He has, however, searched those files to check many a potential solution to problems generated through more than pedagogical or evaluative motives, and he no doubt has observed how explicitly those files contain statements to support the hypothesis that artifacts evolve by the incremental elimination of their defects.

David Pye uses the related example of nuts and bolts to articulate principles relating to the evolution of artifacts:

> When hexagon nuts and hexagon heads superseded the old square ones on bolts, it must have been greater convenience in use which argued for the change: to turn a square nut in an awkward place one may need two different spanners. . . . From that time on for many years hexagon bolts were one of the normal features of "modern engineering." By means of them alone if by nothing else, the layman of the early nineteenth century could distinguish between one of the new engines and the old ones of Watt's time.
>
> Nearly always when a new feature appears it has earned its place by defeating an older one.

But, just as diametrically slotted screw heads have not been totally displaced by Phillips-head ones, so square bolts did not become extinct. There remained applications where the wrench or spanner did not have to fit into an awkward place, and the square nut maintained the advantage of economy over the hexagonal. Erector sets, which sought a high-tech image in the early twentieth century but presented many an awkward space for even a child's hand and toy wrench, still came with slotted bolts and square nuts, as did the contemporary British Meccano sets. Hexagons did have the disadvantage of being too easily rounded into useless circular nuts.

"So long as there are inconveniences and discomforts in our ordinary way of life, so long will there be inventors striving to improve matters." So reads the introduction to a collection of descriptions of short-lived products taken from the patent files of the century beginning in 1849. The pages of *Scientific American,* founded a year earlier; the *Illustrated London News,* dating from the same period; or a host of contemporaneous popular publications could serve just as well to chronicle the drive of inventors. And the catalogues of world's fairs, descendants of the Great Exhibition of the Works of Industry of All Nations, held in London in 1851, provide further evidence of the era of great independent inventors. However, a glance at the "50 and 100 Years Ago" page of any current issue of *Scientific American* shows rather dramatically that perception of things certainly did change, at least in the pages of that magazine, between the early 1890s and the early 1940s. Whereas one hundred years ago there was frequently illustrated a curious gadget or device that improved on existing gadgets and devices, the news of fifty years ago is principally of scientific theories and discoveries, knowledge of which an inventor-manager like Rabinow might find interesting but not necessarily having in itself the makings of a creative invention. By World War II, we seem to have come to take new gadgets for granted or relied upon advertising to inform us of what was new. Whereas our great-grandparents apparently found the latest improvement on the fountain pen or the bicycle of intellectual interest, most people in our generation take only a commercial and utilitarian view of such things. Thus the science-and-technology sections in today's newspapers and magazines will generally print pages of jargon from the medical and physical sciences but give us little of the thoughts or products of engineers or inventors.

Invention is not dead, however, nor is what drives inventors today any different from what did so in years past. The link between the evolution of artifacts and the practice of invention is timeless, even if it has by and large come to take a less visible role in society. What drives today's inventors is the same thing that drove nineteenth-century inventors to put lightning rods on umbrellas or to make the umbrellas integral with headgear so as to free the hands of their wearers.

Whether self-generated or heard from others, whether couched in crass terms of coming up with a million-dollar idea or in utopian dreams of wasteless societies, whether expressed in Anglo-Saxon concretions or in Latinate, polysyllabic abstractions, dissatisfaction with existing artifacts is at the core of all invention and hence all changes in made things. Thus Edwin Land was prompted to invent the Polaroid instant camera by his three-year-old daughter's question as to why she could not see at once a picture he had taken of her. The innocent "Why not?" articulated a shortcoming of existing cameras that Land was determined to remove.

In his classic *History of Mechanical Inventions,* Abbott Payson Usher spoke of the same process of invention in the following, more academic terms:

> Invention finds its distinctive feature in the constructive assimilation of preexisting elements into new syntheses, new patterns, or new configurations of behavior. . . . Invention thus establishes relationships that did not previously exist. In its barest essence, the element of innovation lies in the completion of an incomplete pattern of behavior or in the improvement of a pattern that was unsatisfactory and inadequate.

Seasoned inventors seem clearly to understand and operate under the generalization contained in Usher's remarks, recognizing that it is problems with existing patterns of doing things or problems with the things themselves that provide opportunities for invention—for new, improved patterns.

Whether the new and improved are patented can be a matter of taste or judgment, and some of the most prolific of inventors and engineers—such as Isambard Kingdom Brunel, designer of the Great Western Railway and the *Great Eastern* steamship—have been strong opponents of the patent system, which they felt stifled innovation. In 1851 Brunel wrote to the Select Committee of the House of Lords on the Patent Laws:

> I believe that the most useful and novel inventions and improvements of the present day are mere progressive steps in a highly wrought and highly advanced system suggested by, and dependent on, other previous steps, their whole value and the means of their application probably dependent on the success of some or many other inventions, some old, some new.

Because Brunel believed that "really good improvements are not the result of inspiration," but "more or less the results of an observing mind, brought to bear upon circumstances as they arise," he believed that "most good things are being thought of by many persons at the same time." The patent system obstructs real progress, according to Brunel, because when someone "thinks he has invented something, he immediately dreams of a patent, and of a fortune to be made by it." Brunel continued,

> If he is a rich man he loses his money, and no great harm is done; but if he is a workman, and a poor man, his thoughts are divided between scheming at his machine in secret, and scheming at the mode of raising money to carry it out. He does not consult his fellow-workmen, or men engaged in the same pursuits, as to whether the same thing had ever been tried, why it had failed, what are the difficulties, or (what is most probable) whether something better is not already known, and waiting only the demand.

Other inventors and engineers, such as Henry Bessemer, who was unabashedly interested in fortune, have not deplored the patent system, but have made ad hoc judgments about what to patent. While Bessemer did collect a host of patents throughout his life, notably those protecting his iron-smelting and steel-making processes, he deliberately chose not to patent his method for making bronze powder. He kept this very lucrative business venture secret for about thirty-five years by carrying it out in a secure factory and by employing only trusted relatives in key positions. By Bessemer's own account, the profits "provided the funds demanded by the ceaseless activity" of his "inventive faculties."

While any theory of the evolution of artifacts must be independent of whether the objects of its principles are patented or not, it is obviously easier to find documentation for testing hypotheses in the formal literature of technology than in the trade secrets of family businesses. Indeed, patent files, while by no means a complete record of artifactual evolution, do provide a store of primary sources and case studies. Even the secondary literature of patents and the patent process provides a wealth of insight into the nature of invention and evolution in technology.

Patent It Yourself, a book by patent attorney David Pressman, is more for the "first-time inventor" than the old hand or the theorist, and so it deals on a rather elemental level with much of the process of inventing and patenting. In an early chapter on "the science and magic of inventing," which Pressman admits the experienced or corporate inventor can skip, the process of invention is described as a "two-step procedure: (1) recognizing a problem, and (2) fashioning a solution."

The trick is in taking the first step, which Pressman recognizes can often amount to "about ninety percent of the act of conceiving the invention." His advice to the novice is to ferret out problem areas:

> This can often be done by paying close attention to your daily activities. How do you or others perform tasks? What problems do you encounter and how do you solve them? . . . Ask yourself if something can't be done more easily, cheaply, simply, or reliably, if it can't be made lighter, quicker, stronger, etc.

Later on in his primer, Pressman addresses the question of the commercial promise of an invention conceived to solve a problem, and he advises the reader: "You should not spend significant time or money on your creation until you have thoroughly evaluated it for commercial potential, including considering all of its advantages and disadvantages." He presents a list comprising a "positive and negative factors evaluation," which essentially is a means of establishing whether an invention is indeed a net improvement over what is now being used. In other words, does the invention promise to perform overall better than whatever it is expected to supersede? The problem with such an evaluation is that the positive and negative factors must be weighed, and this involves subjective judgments. For example, Pressman's list contains forty-four factors, ranging from cost, weight, and size to market dependence, difficulty of distribution, and service requirements. The net result of such a survey obviously depends very greatly on how accurately and honestly the relative importance of such disparate factors can be determined. It is clearly not an easy task for a biased inventor to evaluate the fruits of his or her own imagination.

Regardless of the genesis or potential of an invention, the question of "prior art" must be addressed if a patent is sought. This is parlance for whatever knowledge is considered obvious to those most familiar with the area in which a problem is being solved. Thus, to be patentable, an inventor's idea must somehow not be merely an obvious way to improve something. In his book, Jacob Rabinow frequently refers to "the art" in relating stories about how he proceeded with various of his inventions. In one instance in the 1950s, after he had left the National Bureau of Standards to start his own firm, Rabinow was asked by a radio-equipment manufacturer to devise an accurate push-button tuner for an FM radio receiver or a TV set, both of which at the time were relatively new consumer products and were touchy to tune. The inventor had long had an interest in radios and was cognizant of developments in the field. He relates how he was familiar with some of the original tuners, which were as large as the receiver they controlled, and how erratic changes in volume often accompanied their use. He summarizes his expertise with the phrase "I knew the art." Thus Rabinow realized the advantages and disadvantages of existing tuners, which were no doubt obvious to many people working on or with tuners, and he knew how those disadvantages were dealt with. But correcting these faults significantly at any stage in the development of tuners was not so obvious; otherwise, that would have been done for the competitive advantage it would have given to radio and television manufacturers.

By "knowing the art," Rabinow could anticipate, as he worked on the problem of an improved tuner, what features he might incorporate into it that would later make convincing patent claims regarding the prior art. For example, in a departure from the familiar push-button tuner, he devised a pull-button one in which, when pulled out, the conical buttons could also serve as knobs to fine-tune the stations selected. Rabinow's tuner was patented, but the commissioning company did not like the idea of a product's being too different from what other companies were producing: "No one uses pull buttons; everybody uses push buttons." The inventor responded, no doubt precisely because he knew the art, "Well, if you're going to make push buttons, you might as well use the stuff that's on the market. There's nothing I could do that's any better." In other words, he knew the state-of-the-art tuner's shortcomings, but he could not see his way to removing them without striking out in new directions.

For the inventor less experienced than Rabinow, Pressman summarizes the way in which the prior art is treated in a patent application and how the objects and advantages of an invention are conventionally presented to the patent examiners. He explains that the form in an application is first to discuss the prior art, then to present the objects and advantages of the invention, its operation, and finally the claims being made for the invention. Pressman admits that there is a lot of redundancy in a patent application and defends it as effective communication of this kind: "first you tell 'em what you're going to tell 'em, then you tell 'em, then you tell 'em what you've told 'em." And what is told mostly in a patent application is what's wrong with existing things. The patent attorney's advice makes this clear:

> Discuss how the problem to which your invention is directed was approached previously . . . , and then list all the disadvantages of the old ways of doing it. [List again in the objects-and-advantages section] all the things your invention accomplishes and its advantages over the prior art. . . . Include all of the positive factors of your invention . . . and all the disadvantages of the prior art. . . . At the end of this section, add a catch-all paragraph reading as follows: "Further objects and advantages of my invention will become apparent from a consideration of the drawings and ensuing description of it." . . . [And] the objects are effectively repeated again (a third time!) in the concluding paragraph of the specification.

Even if one can endure the tedium of writing out a patent application, there are often even more frustrating aspects to bringing an invention to fruition and commercial success. Although 90 percent of the creative problem of invention may be in problem identification, one does not necessarily have an easy time of it thereafter. An inventor is never home free just because a good problem is identified, for "fashioning a solution," in the sense of conceiving a possible way to alleviate the problem, may involve considerable effort. Thomas Edison was not alone in recognizing problems with candles and gas for lighting, and he conceived but one version of the electric light bulb. (British inventors had long experimented with electric lights before Joseph Swan was granted an English patent for a carbon-filament lamp in 1878, the year before Edison received an American patent for his.) Regardless of priority, Edison's idea for a light bulb was his "inspiration." He then had to test thousands of materials before finding a successful bulb filament to make a practicable model of his idea. Next he had to go through the process of patenting it and, finally, setting up the infrastructure to distribute and sell his invention. Only then was the electric light bulb truly a successful innovation, and it was the long process of going from idea to acceptable product that Edison referred to as the "perspiration" part. Thus when the Wizard of Menlo Park called invention 10 percent inspiration and 90 percent perspiration, he was speaking not only about the creative act of inventing but also about the whole inventive process needed to bring more than intellectual success. Edison warned against discouragement during the perspiration phase in the following way, reminding us that we get things to work by the successive removal of bugs:

> Genius? Nothing! Sticking to it is the genius! Any other bright-minded fellow can accomplish just as much if he will stick like hell and remember nothing that's any good works by itself. You've got to make the damn thing work! . . . I failed my way to success.

While there may be disagreement, as there should be, about what portion of invention and innovation is problem identification, inspiration, or perspiration (Paul MacCready puts Edison's ratio of the latter at 2 to 98 percent, and others put it at 1 to 99), everyone seems to be in agreement that invention begins in identifying a problem in whatever it is that we already have.

Because inventors are inveterate critics of technology, they always see room for improvement, even in the latest and most advanced of artifacts. "The love of improvement," according to the prolific Henry Bessemer, "knows no bounds or finality." Thus the process of technological evolution is a never-ending one, and a study of artifact after artifact reveals that it is the successive identification and elimination of faults found in any given thing at any given time that forms and re-forms its form.

4

From Pins to Paper Clips

Whatever its intended function, an object's form alone often suggests new and more imaginative forms, as the stick did the fork and the shell the spoon. It is no less the case with manufactured things, and few artifacts have been more formed, de-formed, and re-formed than the common paper clip, as a survey once made clear. Attribution of the study and its follow-ups has become as confused as the origins of the object itself, credit going variously to, among others, Lloyd's of London, "relentlessly inquisitive Germans" at a Munich manufacturing firm, and Howard Sufrin, heir to the Pittsburgh family business that made Steel City Gem Paper Clips. According to Sufrin, who claims to have conducted the original study in 1958, three of every ten paper clips were lost, and only one in ten was ever used to hold papers together. Other uses included toothpicks; fingernail and ear cleaners; makeshift fasteners for nylons, bras, and blouses; tie clasps; chips in card games; markers in children's games; decorative chains; and weapons.

I recall the last purpose as the only one to which my classmates and I put paper clips in the early 1950s: We flexed and twisted them apart and used the sharply pointed halves as ammunition for the rubber bands stretched between thumb and index finger. More than one teacher kept the whole class after school when no one would admit to launching the U-shaped missile that had just whizzed past her ear and struck the blackboard—or ricocheted off the ceiling and rung off the waste basket in the corner of the room. We would listen to the familiar lecture on how eyes had been put out by the sharp points of paper clips made into vile projectiles, but we continued to employ these urban slingshots, for none of us had ever actually witnessed a serious injury. The incorrigibles in the class would conduct wars across the back rows, and every time a paper clip pinged off the window the whole class would hold its breath, hoping the teacher would not hear the sound.

Paper clips have also served as objects of more inwardly directed aggression by providing something for the fingers to twist grotesquely out of shape during phone calls, interviews, and meetings. This tactile form of doodling may consume only a fraction of the twenty billion paper clips produced each year, but it underscores the almost limitless functions to which a single form can lead. However paper clips have come to be used or misused, they evolved to their present form only slowly, and at times as circuitously as their wire can be bent. Where to begin the story of something so common in its form and yet so complex in its associations can be as arbitrary and difficult as picking a particular paper clip from a box of a hundred. Just as the clips can get all tangled together, with one pulling others in its wake, so picking up the story of the artifact itself out of the box of cultural and social history inevitably produces a tangle of tales wound around tales.

Paper was developed in first-century China and in time moved westward. By the thirteenth century, making paper from the pulp of linen rags was established in Europe, and the generally available medium on which to write replaced parchment and vellum for all but the most ceremonial and special of documents. In addition to the need for bound volumes of fixed size that were the essentially unchanging records of vital statistics, thought, and achievement, there arose, with the rise of bureaucracy and commerce, increasing amounts of occasional paperwork whose contents did not demand or require rugged or permanent binding. Indeed, it would have been a bother, an expense, and an exercise in pretentiousness to bind two pieces of business paper together as elaborately as were the leaves of books.

Related pages that were not attached often failed to remain together, however. One early way of attaching loose sheets required only a penknife—always close at hand to point quills—and a length of string, a strip of cloth, or a piece of ribbon. Two small parallel slits would be made in the pages to be fastened together with whatever was threaded through the slits, and the ends could be sealed with wax to the paper to ensure that no substitutions were made. In general, the quality of the tie marked the importance of the document, and even today one can find contemporary records fastened in such a manner: I have received from Eastern Europe university

Pin making, shown here in an eighteenth-century print from Diderot's *L'Encyclopédie,* was a classic example of division of labor. The pin, like the needle, was highly evolved long before its manufacture became mechanized.

transcripts whose pages were gathered with exquisite ribbons elaborately tied. But I have also received from underdeveloped countries multipage documents or unofficial copies of records fastened together by another old method—a straight pin.

Pins were fashioned out of iron and bone by the Sumerians as long ago as 3000 B.C., and were used to hold clothes together. The manufacture of pins was industrialized long before it was mechanized, and the manual process was illustrated in Denis Diderot's monumental *L'Encyclopédie,* completed in 1772. In a famous passage near the opening of *Wealth of Nations,* Adam Smith described how a pin was made to demonstrate the advantages of a division of labor: "One man draws the wire, another straightens it, a third cuts it, a fourth points it, a fifth grinds the top for receiving the head. . . ." William Cowper rendered the same process in verse—"One fuses metal o'er the fire; / A second draws it into wire"—thus showing that there is more than one way to make a point.

Wire could be drawn at the rate of sixty feet per minute, but only slightly more than one pin a second could be cut by a practiced worker. This would yield about four thousand pins per hour. The bottleneck in the manufacture of pins occurred when they were attached to cards or papers; the women who worked at that cottage industry accomplished the task at the rate of perhaps fifteen hundred per day. Adam Smith observed that, averaged over all the specialists that divided the labor (and as many as seventeen different people might work on each single pin), about forty-eight hundred pins per day per worker was the output. He speculated that without a division of labor, the output of a single person making each pin from start to finish might be as great as twenty but perhaps as small as a single pin per day.

The efficiency of the division of labor in making pins was a major impediment to mechanizing the industry. But, just as there were many ways in which to divide the hand labor that produced pins, so there would be many ways in which to put together belts and pulleys, cams and gears, shears and hammers, claws and files, to make a pin mechanically. As Steven Lubar, who has written on the cultural as well as the technological influences on design in the pin industry, has warned us, "we must not be misled into thinking that [a pin] machine has the form it has because of some deterministic factors, that physical law requires that pin machines look and operate" alike. One machine was invented and patented as early as 1814 in America, and a more practical one was patented in 1824 in England by an American engineer then living there. But the most successful of the early pin-making machines took its form from the "mechanical scheming" of an erstwhile physician who had watched the process of hand pin-making by the inmates of the New York Alms House, where he was resident.

John Ireland Howe, who was no relation to the inventor of the sewing machine, was born in Ridgefield, Connecticut, in 1793 and began to practice medicine in New York City in 1815. An inventive urge drove him to apply his knowledge of chemistry to produce a practical rubber compound, and after receiving a patent in 1829 he gave up his medical practice to manufacture rubber goods. But when the venture did not succeed, he began to experiment, with an eye toward developing machinery that would replace the many human hands that he had observed making pins in the almshouse. However, he was limited by his lack of mechanical experience and so sought the help of Robert Hoe, a designer and manufacturer of printing presses. It was in Hoe's shop in 1832 that Howe came up with a working model of a machine for making pins in one operation, and it was patented. Although early attempts to sell the imperfect machine were unsuccessful, leading him into considerable debt,

Howe continued successively to remove faults in early models and came up with increasingly improved machines. In 1835 the Howe Manufacturing Company was founded and it soon had five machines operating, with pins being made in both England and America.

At one stage in the development of his business, Howe had three machines, producing seventy-two thousand pins per day, but it took as many as sixty pin stickers to package the output. Thus the true mechanization of the industry required the mechanization of the pin-sticking operation. Eventually Howe and some employees of his company designed a machine to crimp paper into ridges through which the pins could be stuck, and this proved to be a great success. It was a far cry from the situation in the Middle Ages, when pins had become so scarce that a British law allowed pin makers to sell their product only on certain days. "Pin money" was set aside to purchase the dear necessities, but, with mass production and the consequent sharp decline in price, "pin money" came to mean pocket money or "a pittance sufficient to purchase only pins."

The sale of pins on cards came about for several reasons. In the early nineteenth century, people had been used to handmade pins whose quality could vary significantly from piece to piece, some being straighter than others, some having better points than others, some heads being uncomfortably large and others painfully small when holding together (and close to the body) the parts of one's dress. Even after mechanization, by displaying very clearly the head and point of every pin on a card, the manufacturer could demonstrate the uniformly "extra ne plus ultra" quality of the product, and the customer could easily verify that a full count of pins was being bought. Carded pins also were conveniently and safely stored and yet could be at the ready to be picked up when a seamstress might need one in a hurry. A "paper of pins" was a godsend, and to this day pins and needles are packaged in a similar way, even though the advances in mechanization have made the quality of pins very high and reliable.

As the availability of high-quality pins increased and their price dropped, they became available in bulk to the commercial establishments that had grown up with the Industrial Revolution. Although the pins sold to businesses as "bank pins" and to home seamstresses as "toilet pins" (named after the dressing table, not the washroom) were identical in their manufacture, differences in packaging distinguished their price. Bank pins were sold loose in half-pound lots,

Although mechanization was producing highly uniform pins by the middle of the nineteenth century, they continued to be packaged so that the customer could see that there was a full count and that all the heads and points were properly formed. The carding or papering of pins had long been a bottleneck in their production, and the output of the first mechanized pin factories was limited by how fast pins could be put up in such packages.

whereas toilet pins continued to be sold threaded in neat rows through pieces of paper or cardstock, often imprinted with the company's name and claims about the quality of the pins. A card of pins might also contain assorted sizes and types, such as "one row black" for use with darker garments. Commercial purchasers did not require such variety and did not have to be sold on the quality or economy of items that made it possible to attach papers together securely yet quickly for processing. Banknotes could be temporarily attached to invoices for proper crediting and accounting, and then removed, leaving only a couple of tiny pinholes—a distinct advantage over slits large enough for ribbons.

A single bank pin would naturally have been more difficult to pick out of a pile or tray of others, and so they also came to be packaged not on flat cards but in ways that suggested a full pincushion ready for the picking. Some such arrangements are really rolls of long strips of paper, not unlike scrolls, to which a single line of pins is attached sideways, and they are still sold as "pyramids" of pins that can sit at the ready on a clerk's desk, and thus are sometimes called "desk pins." The difficulty of picking up pins from a pile in a desk drawer or tray also led to the evolution of a different form of pin—the "T" pin—which has a large head formed by bending the pin wire sideways and then back upon itself in a tight curve to form a T-shaped head. A current catalogue offering these pins, "used primarily in brokerage houses for securities," really documents the failings of the straight pin that "T" pins overcome: "These pins have handles which speed pick-up, insertion, and withdrawal, will not slip through paper."

By the end of the nineteenth century, pin-making machinery had improved to such a point that a half-pound box of bank pins could be had for forty cents, whereas a much smaller quantity of carded or papered pins for home use sold for about seventy-five cents. Many early pins were made of brass, which is a soft metal and hence not so desirable as steel. Mass production could not keep steel from rusting, however, and so the better pins began to be plated with nickel, but even this metal started to break down and flake off in extremes of humidity, causing rust marks to soil whatever was pinned together.

This shortcoming of steel pins was not especially inconvenient for homemaking uses, where pins were often used only temporarily while sewing or wearing garments, and homeworkers could take care not to leave pins in anything that would be put away for a while. (Pins that did develop rust could be cleaned by pushing them back and forth in a sack of emery grit, which was often sewn into the form of a strawberry.) However, it was necessary in business applications to have piles and files of papers that remained pinned together over long periods of time, and it was impractical to have to worry about them or clean the pins of rust. Another disadvantage of using pins to attach business papers together was the unsightly holes, often ringed with rust, that remained. This was an especially bothersome problem when papers were attached, detached, and reattached time after time over the course of years. The pinned corners began to become rather ragged, and so alternatives that corrected this flaw were sought.

To obviate the undesirable use of pins to attach papers together, inventors as early as the middle of the nineteenth century developed what were called "paper fasteners" and "paper clips," though this latter term at first designated the kinds of bulky spring devices that we today find on clipboards. Among the first of smaller paper fasteners to be patented was a decorative metal device whose two small teeth pierced the papers and were folded over another piece of metal placed against the back side of the sheets, thus clasping them together. Though this did not remove the problem of the papers' having holes made in them, it did reduce the tendency of a sharp point to snag other papers on the desk. A greater promise of the new fasteners was protection from pricked fingers when shuffling papers. According to the 1864 patent, the new fasteners also overcame the failings of other means of fastening papers: "The corners of the sheets are completely prevented from turning over or being disfigured with 'dog ears,' as is usually the case with legal manuscripts."

A great variety of such fasteners came into existence in the last quarter of the nineteenth century, and there was fierce competition among them. As in the evolution of all artifacts, each variation of a fastener promised to solve some or all of the problems of the preexisting forms. One style, the Premier fastener, advertised that its points did not become crushed "as in fasteners similar in appearance to the Premier." Fasteners of a dissimilar appearance were also developed to answer the objection to the paper-piercing points altogether. One class of such devices was patented by Ethelbert Middleton of Philadelphia in 1887, and his "improvements in paper fasteners" consisted of malleable metal stamped in curious patterns whose use involved the action of folding various wings over the corners of the papers, which "effectually secures the clasp in position on the mass of papers without any puncturing or cutting of the papers themselves." Variations of all these paper fasteners, both piercing and folding, are still made and sold today, for those who might prefer their papers pierced or crimped in the corner to having them slipping apart.

While Middleton's improvement did not pierce papers, it did hold them securely by gripping with the convoluted edges on its bottom flap. However, the multistep process of folding various wings over to attach the fastener was not an attractive feature. Any single device that would deal with both problems, eliminating both paper piercing and complexity of attachment and removal, would have a distinct advantage. Mass production of mid-to-late-nineteenth-century paper fasteners had been made possible by machinery that could stamp out of sheet metal great quantities of products quickly and effectively. In the last quarter of the nineteenth century, new machines appeared that were capable of bending and shaping objects from spring-steel wire. These descendants of pin-making machines enabled an entirely new form of paper fastener to be developed in response to the shortcomings of existing ones.

A successful paper clip begins with steel wire that wants to spring back to its original shape after being bent, but only up to a point, for otherwise the paper clip could not be formed into the clever and pleasant-looking object that it is. Steel and all materials are said to behave "elastically" when they stretch, bend, or twist in proportion to the force applied to them and resume their original shape after being let go, as observed by the English physicist and inventor Robert Hooke, who discovered in 1660 the principle that now bears his name. He did not publish it until 1678, however. Even then, in the manner of his times, which included fierce competitiveness over claims to priority, Hooke did not actually articulate the principle but merely published it in the form of the Latin anagram *ceiiinosssttuu.* Two years later, when he was so inclined, he rearranged the letters into the phrase *Ut tensio sic uis* and explained that "as the tension so the power" of a spring meant that the more you pull the more it resists—until you pull too hard and the spring gives and does not completely return to its original shape.

Forming a paper clip presents a common dilemma encountered by engineers and inventors: the very properties of the material that make it possible to be shaped into a useful object also limit its use. If one were to try to make a paper clip out of wire that stayed bent too easily, it would have little spring and not hold papers very tightly. On the other hand, if one were to use wire that did not stay bent, then the clip could not even be formed. Thus, understanding the fundamental behavior of materials and how to employ them to advantage is often a principal reason that something as seemingly simple as a paper clip cannot be developed sooner than it is.

Steel wire was still new in the second half of the nineteenth century, and early wire manufacturers looked for applications of their product. Some, like John Roebling, went so far as to promote, design, and build suspension bridges, which used large quantities of wire in their cables. (The elastic springiness of large bridges is often very noticeable to motorists stopped in traffic upon them. If, in the process of erection or use, the steel cables were stretched beyond the limits of Hooke's Law, the bridge would sag permanently like a melted plastic model of itself.) But, whether spinning bridge cables or bending wire into fasteners, specialized machinery was essential for exploiting the new material. To have formed paper clips one by one by hand would have made them very expensive and hardly a challenge in business applications to the modest machine-made straight pin. Hence the widespread manufacture and use of the paper clip had to await not only the availability of the right wire but also the existence of machinery capable of tirelessly and reliably bending it in a flash into things that could be bought for pennies a box. In the meantime, though there may not have been an outcry of complaints about desk pins, there no doubt were numerous inventors and would-be inventors who found the pin an unsightly and inappropriate paper fastener and thought there must be a better way.

As with many new devices, especially ones of modest proportions and few pretensions, the origins of the first bent-wire paper clip are not without their uncertainties, including those induced by chauvinism. According to an oft-repeated account, a Norwegian named Johan Vaaler should be credited with the invention of the paper clip in 1899. However, as the story goes, Norway had no patent law at the time, and though Vaaler's drawing was accepted by a special government commission, he had to seek an actual patent in Germany. Norwegians are said to have remembered proudly the humble item's origins in their country when, during World War II, they "fastened paper clips to their jacket lapels to show patriotism and irritate the Germans." Wearing a paper clip could result in arrest, but the function of the device, "to bind together," took on the fiercely symbolic meaning of "people joining against the forces of occupation."

Vaaler's *fin-de-siècle* notion was granted an American patent in 1901, and the document describes the "paper clip or holder":

> It consists of forming same of a spring material, such as a piece of wire, that is bent to a rectangular, triangular, or otherwise shaped hoop, the end parts of which wire piece form members or tongues lying side by side in contrary directions.

As if to emphasize that a paper clip need take no unique shape, several styles are illustrated in Vaaler's patent. (Such a multitude of

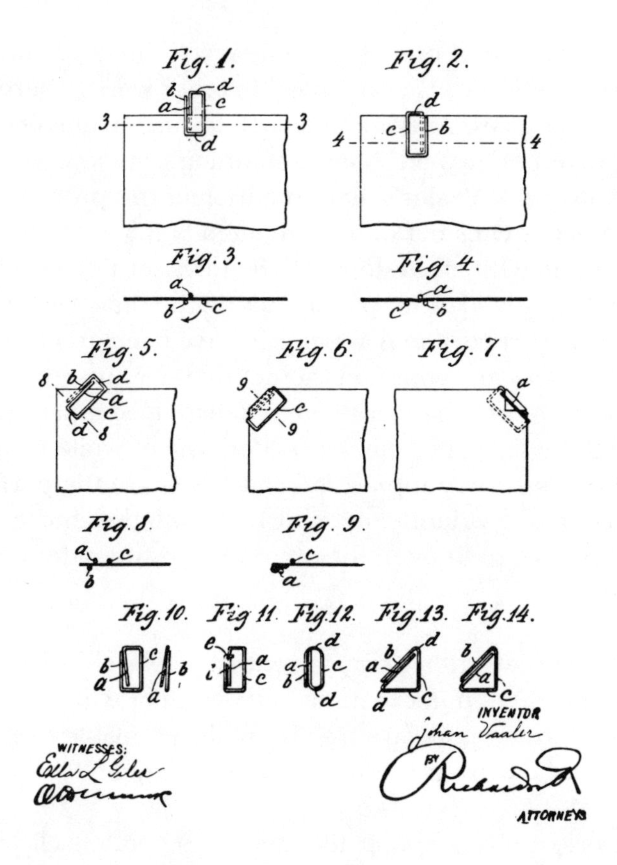

Johan Vaaler's first American patent, dated June 4, 1901, showed several embodiments of a "paper clip or holder." The version labeled "Fig. 12" suggests the beginnings of what has come to be known as the Gem paper clip, but is clearly not a fully formed Gem.

ways to achieve the same end is common in patent applications, which thus provide counterexample after counterexample to the claim that form follows function.) Even if Vaaler's paper clips look superficially like today's, they differ in one major respect: the wire does not form a loop within a loop. Papers would be held together by the arms of the clips, of course, but they would have required a very deliberate action to apply. Curiously for such simple devices, neither Vaaler's nor most other contemporary patent applications for paper clips included a model.

Vaaler did make explicit a secondary feature of his invention: "To obviate the clips hanging together when being packed up in boxes or the like, the end of one of the tongues . . . may . . . lie close up to the base part of the other tongue." In other words, there were no great protuberances. Anticipating such an inconvenience as clips' "hanging together" would show uncommon prescience on the part of an inventor, and Vaaler's mere mention of this problem suggests that other paper clips not only were already in existence but were annoying users with an undesirable feature not unheard of today.

There were indeed other paper clips by the time Vaaler's American patent was issued, and it appears to have been granted more for his variations on some common themes than for any seminal contribution. Matthew Schooley, a Pennsylvanian, filed a patent application in 1896 for a "paper clip or holder which, while simple in its construction, is easy of application and certain in the performance of its functions." Evidently, even at that time the shortcomings of such a device were known, for according to the patent, issued in 1898:

> I am aware that prior to my invention paper-clips have been made somewhat similar to mine in their general idea; but so far as I am informed none are free from objectionable projections which stand out from the papers which they hold.

Furthermore, unlike a Vaaler-like design, Schooley's clip would lie "flat upon or against the papers which it binds together [without] puckering or bending" them. It achieved this by overlapping the wire in coil fashion. Though there is no turn within a turn, its form suggests today's paper clip almost as much as Vaaler's does.

When all is said and done, any attempt to sort out the origins and the patent history of the paper clip may be an exercise in futility. For there appear to have been countless variations on the device, a great multiplicity of forms, and some of the earliest and most interesting versions seem not to have been patented at all, which is perhaps not so surprising for such a modest artifact. Nevertheless, however obscure their provenance, there is little doubt that alternate forms of the artifact evolved in response to the failure of existing forms to reach perfection, and therein lies the value of this most common object as a case study of how failure can drive form to fanciful extremes in quest of parallel objectives.

In 1900 an American patent was issued to Cornelius Brosnan of Springfield, Massachusetts, for a "paper-clip," which has been re-

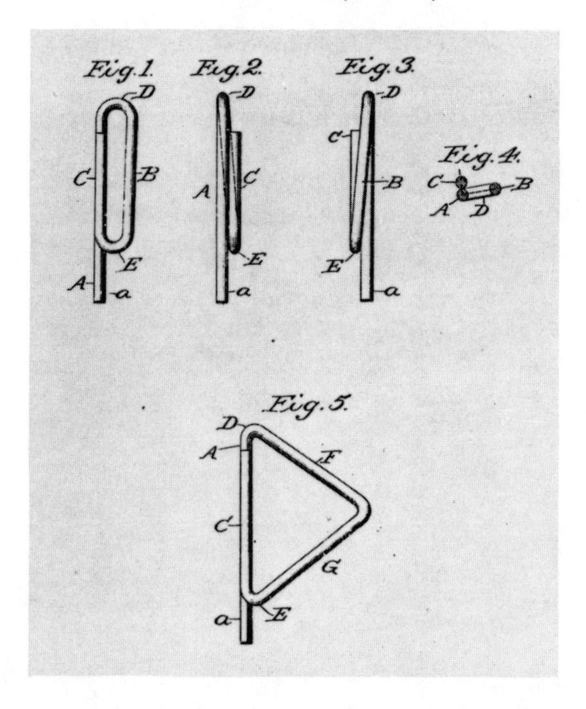

An American patent for a bent-wire "paper clip or holder" was issued in 1898 to Matthew Schooley, thus predating the commonly cited "invention" of the paper clip in 1899 by the Norwegian Johan Vaaler. Just as Schooley's patent drawings show different embodiments of the clip, so there are believed to have existed many other (unpatented) variations, some dating from as early as the 1870s.

garded in the industry as the "first successful bent wire paper clip." Again, no model was submitted, but two versions of the clip were shown in the patent drawings, and they are evocative of the outlines of track layouts that graced the Lionel and American Flyer model-train catalogues that I pored over as a child. Nevertheless, in the typical manner of patent literature, Brosnan's description of "certain new and useful Improvements in Paper-Clips" suggested the problems with existing ways of fastening papers that the new clip would overcome:

> This invention relates to an improved clip or binder for fastening together sheets of paper, the object being to provide a form of paper fastener which may be rapidly and cheaply made in large quantities from spring-wire, having the capability of being

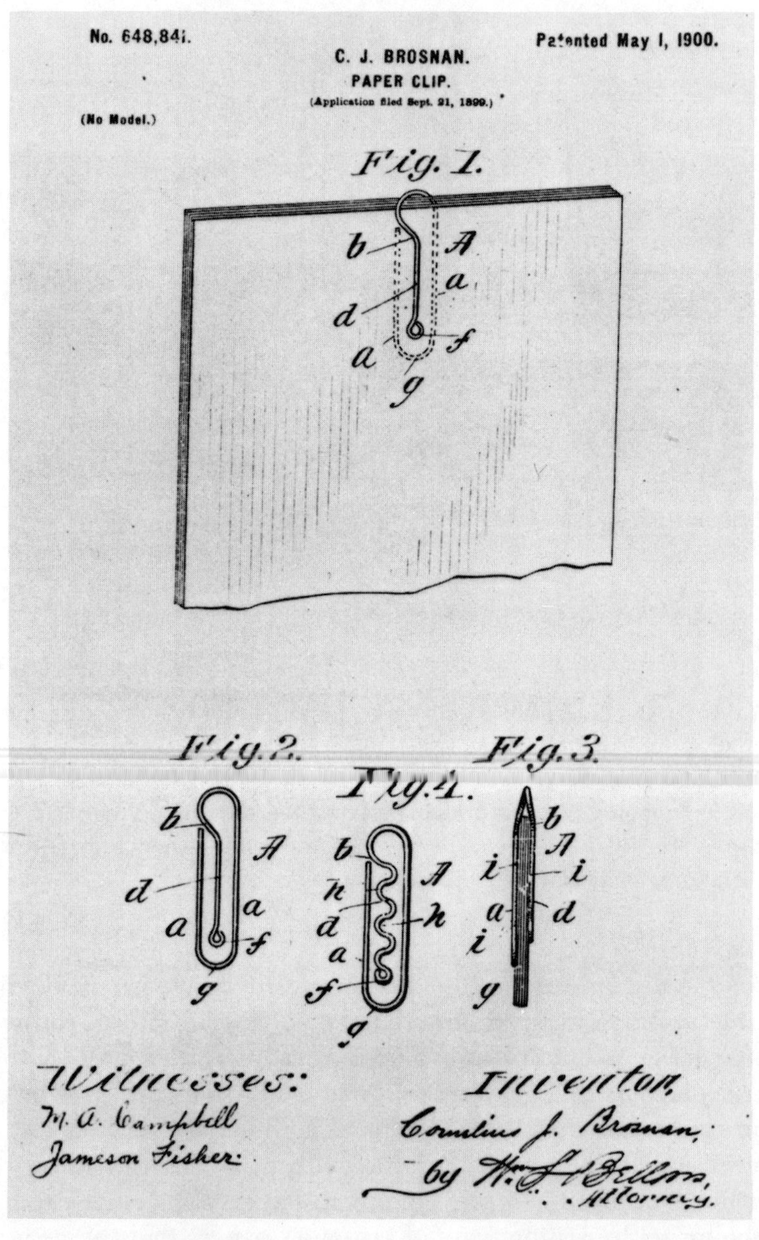

In 1900, Cornelius Brosnan was issued a patent for a paper clip that removed one of the principal objections to many earlier designs. Because the inner leg of this Konaclip terminated in a tight loop, or "eye," it did not have a sharp end to catch, scratch, or tear the papers it held. However, Brosnan implicitly admitted a failing of his own Konaclip by taking out another patent five years later for a clip that did not have an eye to hook in the loop of a box mate.

> very conveniently applied in its fastening engagement with several sheets of paper, of holding the papers together with all required security, and yet of permitting its disengagement when desired.

Clearly, at least in the minds of the inventor Brosnan and the patent examiner, the new paper clip was superior to existing devices, and its unique form was described in three separate claims, each of which began:

> A clip or paper-fastener constructed of a single length of wire bent to form an elongated frame with an end portion of the wire deflected inwardly within and near one end of the frame, and extended longitudinally along and within the middle of the device . . .

The claims went on to specify that the wire was "formed corrugated, and terminating in an eye . . . near the other end of the frame." This eye kept the clip from scratching or tearing the papers it was attached to, something Schooley's and Vaaler's clips were always liable to do. Brosnan did have something: his clip, which was called the Konaclip, took full advantage of the latest technology to bend wire into tight loops, and far surpassed anything then patented. It was certainly easier to apply than most other existing designs. Even so, the Konaclip did not last long, for, notwithstanding his promise that papers would not slip from it, they did, especially the middle ones in a pile.

Brosnan, like many other inventors, no doubt thought he had spelled out his claims in such a way as to cover all practical methods of bending a piece of wire to serve as a "perfected" paper clip. But perhaps nothing mocks the cliché that "form follows function" so emphatically as this common object. The eye formed at one end of the Konaclip seemed essential, for example, because if the inside of the clip had terminated instead in a straight piece of wire, this could have snagged and pierced papers as the clip was attached, thus reducing any advantage over the pin. But in spite of Brosnan's advertising claim that his clip, as the final result of a long process of improvement, provided "the only satisfactory attachment of papers," and his warning to businesspeople, "Don't mutilate your papers with pins or fasteners," the fact remained that papers still slipped out from paper clips. Besides, Brosnan's designs were sure to snag one another in the box.

In 1905 Brosnan was issued a new patent for a paper clip "of novel shape" that was "formed of spring-wire and constructed with portions which [when] distended oppositely from each other develop a reaction to embrace and bind marginally sheets of paper." This clip did not depend upon overlapping wire for its gripping power, for, like the Konaclip, it was formed all in a single plane. Rather, the papers were grasped by the spring action created by separating the touching inner and outer loops of wire. According to Brosnan's patent, his new clip had the advantages of "cheapness of construction, ease of manipulation . . . , efficiency in the retention and binding action . . . without liability of swinging or shifting from its given set position, and . . . not becom[ing] interlocked one with another to cause bother and delay in taking one or more from the box . . . , [and not catching] other papers in conjunction with which the clipped bunch of papers may be brought." It is clear what the disadvantages and failings of existing paper clips were.

Several of the many alternative forms to the paper clips that Brosnan and other ingenious wire benders came up with are recorded in the pages of *Webster's New International Dictionary*. As if to emphasize the importance of the form of a paper clip and the difficulty of defining it in words alone, the definition is illustrated. The first edition, dating from 1909, defined "clip" as "a clasp or holder for letters, bills, clippings, etc.," and showed a predecessor clamplike device along with Brosnan's Konaclip and some alternative ways to bend wire that he had not anticipated in his patent claims. These clips, which have come to be known as the Niagara and Rinklip, demonstrated, for example, that one did not need an eye or a loop terminating within a wire frame to make the thing function. When the second edition of *Webster's* was published, in 1934, "paper clip" was defined as "a device consisting of a length of wire bent into flat loops that can be separated by a slight pressure to clasp several sheets of paper together." The reader was referred to the "clip" entry for an illustration that no longer included a stamped-metal style or Brosnan's Konaclip, but showed still another way of forming a clip—with two "eyes" outside the main body of the clip. This design was to evolve into one with its eyes inside the wire frame, where they were less likely to snag their box mates, a design

Illust. **b** *Angling.* A gaff or hook for use in landing the fish, as in salmon or trout fishing. *Scot. & Dial. Eng.* **c** A grappling iron. **d** A clasp or holder for letters, bills, clippings, etc. **e** An embracing strap, as of iron or brass, for connecting parts together; specif., the iron strap, with loop, at either end of a whiffletree. **f** Any of various devices for confining the bottom of a trousers leg, used in bicycling. **g** A device to hold several, usually five, cartridges for charging the maga-

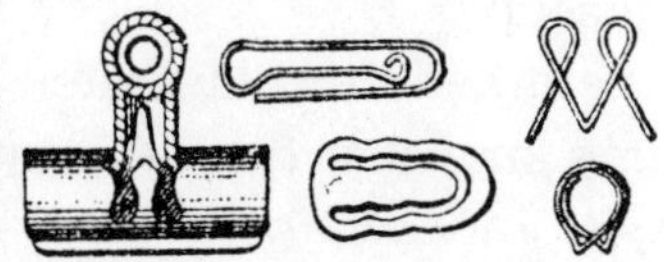

Various forms of Clips for papers.

2. That which clips, or clasps; a device for clasping and holding tightly, as: **a** A grappling iron. **b** A clasp or holder for letters, bills, clippings, etc. **c** An embracing strap, as of iron or brass, for connecting parts together; specif., the iron strap, with loop, at either end of a whiffletree. **d** Any of various devices for confining the bottom of a trousers leg, used in bicycling. **e** *Scot. & Dial. Eng.* An instrument for lifting pots, etc., from a fire, or for carrying barrels, etc.

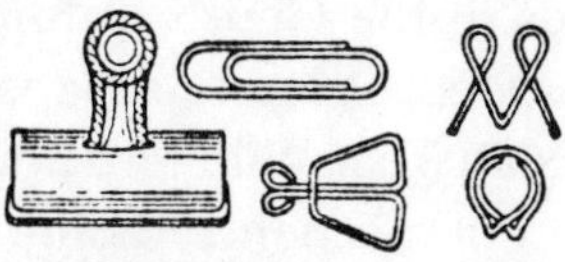

Various forms of Clips for papers.

The illustration accompanying the definition of "clip" in the first edition of *Webster's New International Dictionary* (1909, top) included a device stamped out of sheet metal and a Konaclip. In the second edition (1934, bottom), the sheet-metal clip was replaced by an early version of an Owl-style clip and the Konaclip was replaced by a Gem. The paper clamp, as well as the Niagara (extreme right, top) and Rinklip (extreme right, bottom), both of which are easily applied to papers, retained their specialized positions among a growing variety of paper clips.

that would come to be known as the Owl style. In early advertisements its superiority to the one-eyed Konaclip was proclaimed in verse:

An eye for business
One too few
Observe this clip
This clip has two.

The advantages claimed for the Owl clip included—besides that of not tending to get tangled with others of its kind—the absence of any sharp ends that might snatch at papers which did not belong with the clipped pack, or might rip papers upon removal. As matters turned out, however, it was not the Owl that was destined to drive out the Konaclip.

One of the other clips illustrated in the second edition of *Webster's* became known as the Gem, and has now long been the most popular style. Indeed, to the vast majority of people today it is virtually synonymous with paper clip.

The Gem paper clip did not just quietly develop between the first and second editions of *Webster's Unabridged,* however. Indeed, in spite of the Norwegian claim to have produced the archetypal paper clip, the idea of the Gem was fully formed by the time of Vaaler's patent. It existed, at least on paper, as early as April 27, 1899, for it was on that date that William Middlebrook of Waterbury, Connecticut—a center of mechanized pin-making—filed a patent application for a "machine for making paper clips" and showed a perfectly proportioned Gem as the product of the machine. Since Middlebrook patented only the machine and not the clip itself, the Gem design may have predated his application and been already known to those practiced in the art.

Even though the paper clip shown in Vaaler's 1901 U.S. patent was not nearly so fully developed as the one pictured with Middlebrook's machine, American manufacturers have remained cautious about making direct claims concerning the role of their predecessors in inventing the Gem. An anonymous history that appeared in a 1975 issue of *Office Products* does describe Brosnan's 1900 patent Konaclip as a "direct ancestor of the Gem pattern," but Middlebrook's 1899 patent shows clearly that the ancestry was at best reversed. Another history—published in 1973 by a Smithsonian Institution staff member who would not admit to being a curator of paper clips but only to being "merely their protector"—states that no patented clip was "overwhelmingly successful until the beginnings of the 20th century when the *Gem* pattern paper clip was introduced." The language is understandably ambiguous and vague as to the Gem's patent status and nationality, and this case study points out clearly the limitations of relying entirely upon the patent literature for tracing the evolution of artifacts. A search through U.S. patents for paper clips alone will never turn up an unadulterated Gem, and Middlebrook's patent for a *machine* for making them might easily be passed over as not directly relevant to the form of the artifact being manufactured.

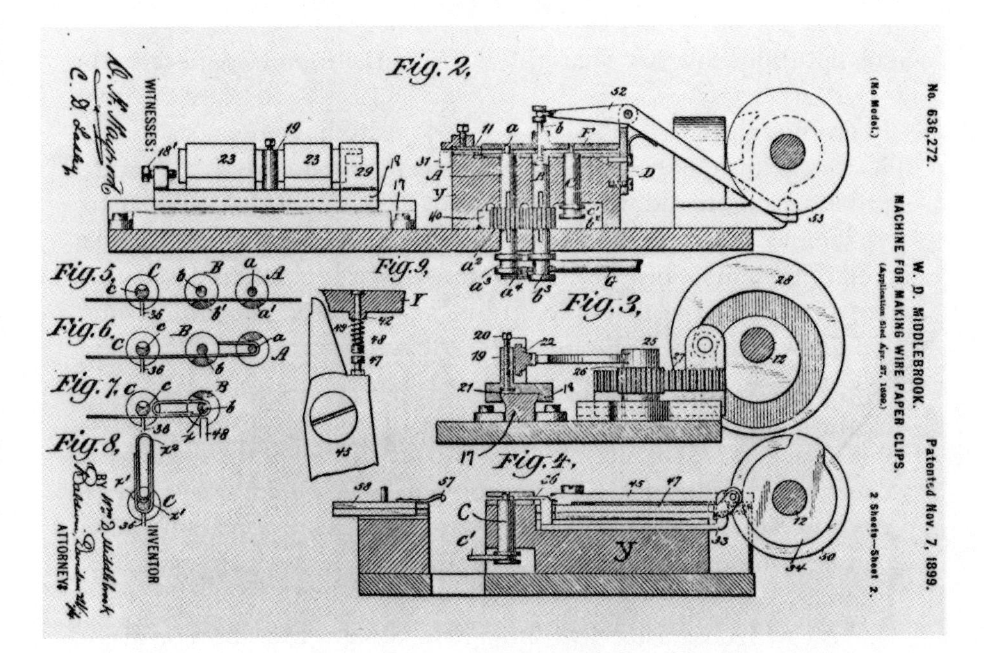

Although the 1899 patent issued to William Middlebrook of Waterbury, Connecticut, was for a machine for making wire paper clips rather than for the clip design itself, Middlebrook's drawings showed clearly (especially in his Fig. 8) that what came to be known as a Gem was being formed. This style of paper clip, which seems never to have been explicitly patented, came to be the standard to be improved upon. While functionally as deficient as myriad other styles, its aesthetic qualities appear to have raised it to the status of artifactual icon.

The Gem paper clip appears to have had its real origin in Great Britain, and the name is said by one international firm to have been "derived from the original parent company, Gem Limited." This is supported by the Army and Navy Co-operative Society's 1907 catalogue of the "very best English goods," which pictures only one style of modern paper clip—a perfectly proportioned Gem, which is described as the "slide on" paper clip that "will hold securely your letters, documents or memoranda without perforation or mutilation until you wish to release them." As early as 1908, the clip was being advertised in America as the "most popular clip" and "the only satisfactory device for temporary attachment of papers." The ad copy went on to warn paper clip users against the use of other existing devices, whose shortcomings the

Gem naturally did not share: "Don't mutilate your papers with pins or fasteners."

Even though the Gem itself never seems to have been patented in its classic form, nor to have been so perfectly functioning a paper clip that inventors did not try to improve upon it, it does appear to have long ago won the hearts and minds of designers and critics as the epitome of possible solutions to the design problem of fastening papers together. For example, in his book *Elegant Solutions,* subtitled *Quintessential Technology for a User-friendly World,* Owen Edwards describes the Gem as follows:

> If all that survives of our fatally flawed civilization is the humble paper clip, archaeologists from some galaxy far, far away may give us more credit than we deserve. In our vast catalog of material innovation, no more perfectly conceived object exists. . . .
>
> With its bravura loop-within-a-loop design, the clip corrals the most chaotic paper simply by obeying Hooke's law. . . .

The Gem certainly has a pleasing form, at least before it has been used and its loops have been misshapen into what looks like a roller coaster, but that pristine form alone seems all too often to have dazzled industrial designers and critics into thinking it works better than it does. Paul Goldberger, for example, in celebrating the design of some common objects, has written:

> Could there possibly be anything better than a paper clip to do the job that a paper clip does? The common paper clip is light, inexpensive, strong, easy to use, and quite good-looking. There is a neatness of line to it that could not violate the ethos of any purist. One could not really improve on the paper clip, and the innumerable attempts to try—such as awkward, larger plastic clips in various colors, or paper clips with square instead of rounded ends—only underscore the quality of the real thing.

One gets the impression that it is certainly the Gem that Goldberger has in mind as the "real thing," and the illustration accompanying his essay confirms this. Few but inventors would argue with the qualities that are ascribed to the paper clip, and many would agree that the newer plastic clips are not just awkward but downright antifunctional (although their nonmagnetic quality may be invaluable for certain computer applications). But many inventors, and not a few users, have disagreed with the idea that "One could not really improve on the paper clip." The "clips with square instead of rounded ends," for example, were considered a distinct improvement by their inventor, Henry Lankenau, of Verona, New Jersey, whose patent is dated December 25, 1934. In typical fashion, but untypically naming the competition, he spelled out his device's advantages in comparison to the failings of existing devices:

> An object of this invention is to provide a paper clip . . . one end portion of the clip consisting of a single loop of rectangular form and the opposite end portion consisting of a double loop and being V-shaped in lengthwise direction.
>
> Another object of this invention is to provide a spring wire clip having two spaced V-shaped loops at one end, the said end providing a wedge action and being adapted to be more easily applied to two or more papers than the type of clip generally known in the art as "Gem" clips and having U-shaped loops.
>
> Another object of this invention is to provide a spring wire clip having a rectangular end portion, the two ends of the wire terminating in a plane lying substantially in abutment with the said rectangular end so as to provide maximum gripping surface and prevent the ends of the spring wire from digging into the papers to which the clip is attached.

Lankenau reiterated this last advantage in the course of his descriptions of the figures depicting several variations of his rectangular-ended clip. In particular, he pointed out that by being set close to the end of the clip the free ends of the wire "cannot dig in and scratch the paper as is usually the case when removing paper clips of the 'Gem' type having short legs which do not extend to the extreme end of the clip." He is right in his criticism of the Gem, of course, and it is possibly because the Gem's classic lines would be ruined by extending the ends of the wire to minimize their digging and scratching that the change has not been made. Lankenau's paper clip, which seemed to deliver on its promise to perfect the Gem, came to be sold as the Perfect Gem, but it is generically known as a Gothic-style paper clip to contrast its features with the Gem's Romanesque appearance. Some self-conscious users, such as librari-

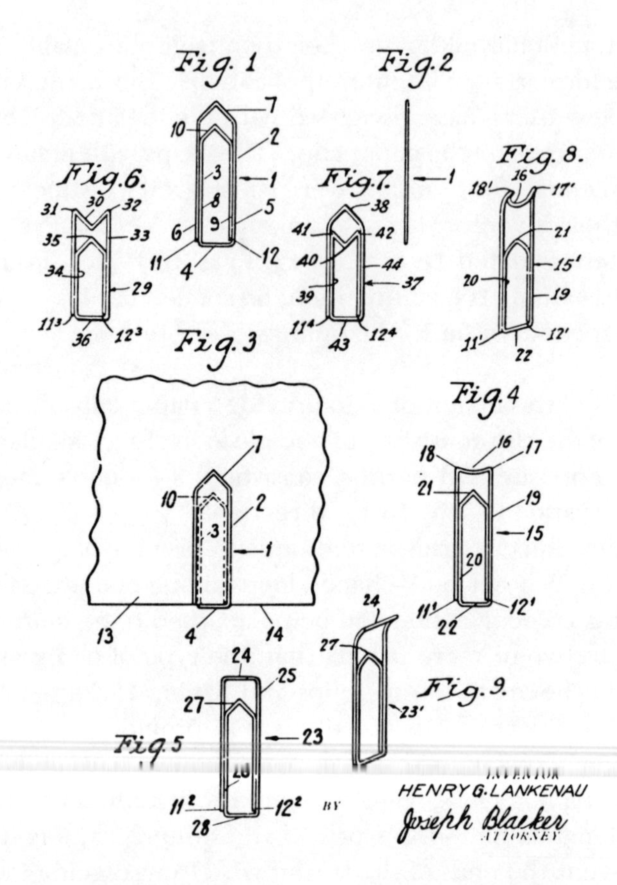

The inventor Henry Lankenau found that Gem paper clips left something to be desired, in that their rounded shape made it hard to get them started onto a group of papers. Some of his sharp-angled designs, patented in 1934, even had the end of the clip bent out of its plane, to provide easier access to the papers—a feature of some paper clips today.

ans who must attach cataloguing material to title pages in the course of processing books, swear that the Gothic clip is much less likely to do damage upon being removed.

Lankenau's patent for the Gothic clip was assigned to the Noesting Pin Ticket Company, then of Mount Vernon, New York. This company was founded in 1913 to make a new kind of pin ticket—a size tag that contained integral pins for attachment to garments. Conventional sharp-pointed pin tickets were notorious not only for doing damage to the clothes they marked but also for pricking the fingers of salespeople and customers. The Noesting Company took its name from the new pin ticket that had rounded bent-wire ends and hence held "no sting" in store. Since the company had the wire-bending capability to make its patented pin tickets, it looked to making other products that required similar wire forming. Paper clips were a natural, and the company now claims to have made "the world's largest selection of paper clips for over 75 years." Visitors to the 1939 New York World's Fair were invited to visit the company's world headquarters and factory in the Bronx, just across the Triborough Bridge from the fair site in Flushing Meadows.

The paper-clip pages of Noesting's 1989 catalogue are a primer on the convoluted relation between the form and function—or, rather, functions—of even so seemingly simple an artifact as a cleverly bent piece of wire. Each different style of paper clip has some advantages over the others, of course, and no single form would appear capable of helping one corral successfully all the chaotic paper on one's desk. Although arranged more in order of popularity than of chronology of development, the clips in Noesting's catalogue are described by their relative advantages, which necessarily imply the disadvantages, shortcomings, and failings of the others. The familiar Gem comes first, in three sizes but without further description or qualification. (Its reputation precedes it!) The Gem is followed by the "frictioned Gem," which has small incisions or notches cut across its length to provide "more gripping power than our standard Gems." Next come the Perfect Gems, whose "patented design makes putting clips on paper easier," and then the Marcel Gems, whose "corrugated surface provides maximum gripping power." The best features these most popular paper clips possess individually are combined in the spread-legged Universal (also known as the Imperial) Clip, whose "unique design . . . allows for easy application with tremendous gripping power."

As we all know, putting even the best-looking of paper clips on cards can be tricky and, once achieved, makes a pile of them awfully bulky. Thus the Nifty Clip was "designed for holding thicker grades of papers such as card or index stock [and is] flattened to conserve card file space." The Peerless (Owl) Clip, whose "rounded eyes prevent catching and tearing," not only "holds more than Gems" but with "greater tension than Gems." Ring Clips, essentially copies of the old Rinklips, are "used when holding only a few sheets," come in five sizes, and possess the advantages of having "less thickness than Gems" and using "less space in files." The last clip offered on the page is the Glide-on Clip, which provides a "tighter grip than Gems when holding small amounts of paper." Clearly the Gem is the standard against which all others are compared, and the comparisons can be made successfully because, for all its "perfection" of form, the Gem does not function perfectly in every situation. Not surprisingly, it cannot be all things to all papers.

The Noesting catalogue also offers "precious metal products," which consist of paper clips "designed for the discriminating executive who wants to make a statement" through the products selected for the "office environment." Furthermore, these clips "allow for specialty applications where standard products do not function properly." Among the items offered are gold-plated Gems, which "will *never* tarnish or rust" and which provide "an ice-breaker for prospective clients." They are "at home on mahogany desks and in boardrooms, yet can add a dash of flavor and class to even the most frugal office." For the more (or less?) frugal office, there are stainless-steel clips, which have the distinct advantages of being nonmagnetic ("safe for use with diskettes"), extremely strong ("powerful gripping power"), and rust-proof ("perfect for archives, law firms, libraries") There are also brass-plated models, "ideal when a gold-tone clip is desired at a more economical price." These can be Gems, Marcel Gems, and Nifty Clips. The last, known also as Ideals, are the large angular clips that look like origami in steel, and they are sometimes called paper "clamps" because they come in sizes capable of holding as much as two inches of papers together rather effectively, something even the jumbo Gem fails to do.

There are still other styles of paper clips offered by other companies with wire-bending experience and machinery, and the variety reminds us not only of the nonuniqueness of form for this object but also of the fact that nontechnological (and subjective) factors such as aesthetics can account for the competitive dominance of one particular form over functionally superior forms. The technological capability to mass-produce bent-wire products was essential in the process of displacing the straight pin by the paper clip; that same capability has provided the proliferation of forms that paper clips have taken. The forms that have survived and thrived have done so in part because of their economical use of wire, but that alone does not ensure success. The Queen City clip, perhaps the simplest and

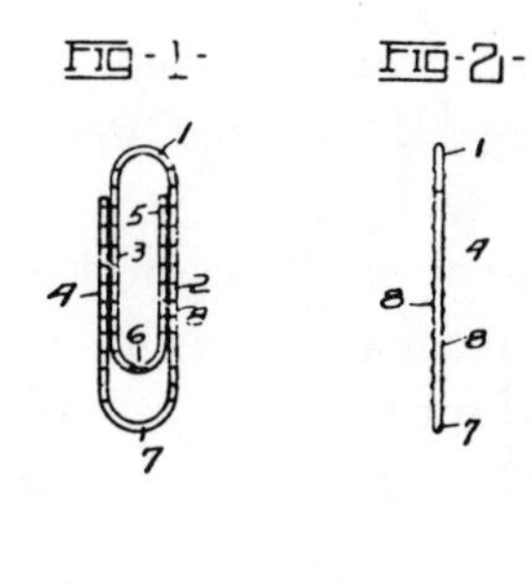

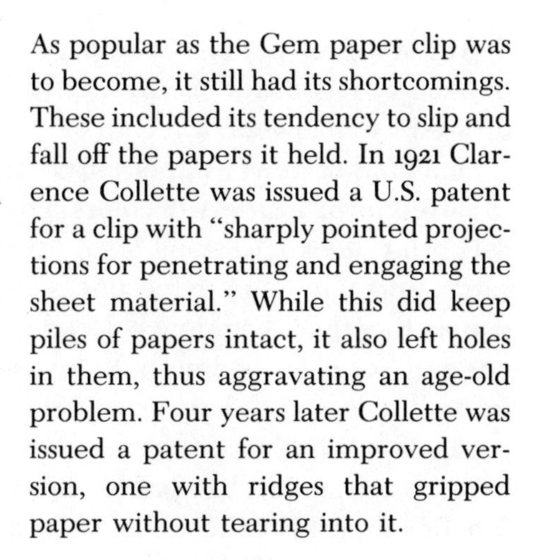
As popular as the Gem paper clip was to become, it still had its shortcomings. These included its tendency to slip and fall off the papers it held. In 1921 Clarence Collette was issued a U.S. patent for a clip with "sharply pointed projections for penetrating and engaging the sheet material." While this did keep piles of papers intact, it also left holes in them, thus aggravating an age-old problem. Four years later Collette was issued a patent for an improved version, one with ridges that gripped paper without tearing into it.

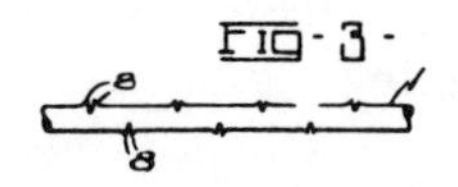

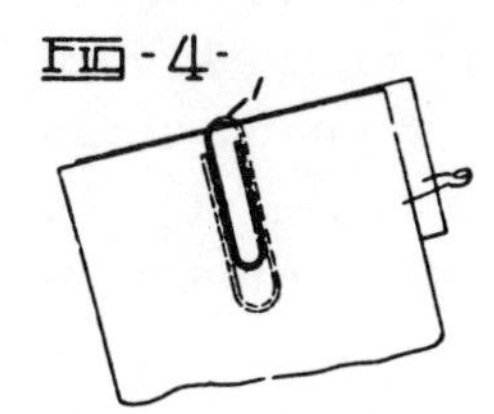

least expensive of designs, has neither the finished appearance of the Gem nor its functional success. Though the Gem is not so functionally perfect as industrial designers might wish, it is the compromise in form of aesthetics, economics, and function that has been embraced by an overwhelming consensus of (technologically uncritical) critics and users alike. It has thus become a standard approached with difficulty by even functionally superior forms.

The ultimate form of the paper clip, whether embodied in the Romanesque, the Gothic, or the oddity, seemed to have become well established by the 1930s, and it has remained virtually unchallenged in the marketplace for half a century, although not because it has ceased to be a challenge to inventors. As late as 1962, Howard Sufrin could say of the firm that manufactured Steel City Gems, "We average ten letters a month from people who think they have an improvement." All such suggestions for changing size, color, and shape may now seem futile, however, for the Gem has long been raised to design icon, and its grip on the minds of critics is no doubt more secure than its grip on their manuscripts. But of late some newer styles of paper clips have become more visible, and their popularity has introduced another complication that must be addressed by followers of form.

One kind of newer paper clip is made of plastic-coated wire and so can come in a variety of colors. Though color-coded clips of folded flat-spring stock have long been utilized for marking records, notecards, and files, they have not generally been used for fastening papers together. The new colored paper clips seem to be intended not only for color coding but also as a means of adding some color to drab offices and dry correspondence, or so it seems from their packaging. Whether or not these are desirable or legitimate ends to which bosses might like paper clips to be put, my experience of the functional performance of at least some of these clips has been less than satisfactory. Their rubbery plastic coating gives them a much higher coefficient of friction than metal and thus can make it literally an effort not unlike pushing an eraser to attach them to a group of papers, which can be wrinkled beyond reason in the process. Furthermore, perhaps because their wire is so much thinner to accommodate the plastic coating without making the clip seem malproportioned, they appear to bend out of shape much more easily than bare metal clips. Why these paper clips have gained such widespread popularity is a functional mystery but a fine example of the role aesthetics and style can play in the evolution of artifacts. Yet this is at the same time but another manifestation of form following failure, for newer, brighter models sell only because the older models fail to be perceived by some users as stylish.

Fully plastic (and colored) paper clips were introduced in the 1950s, but never gained much popularity. These are normally of a roughly triangular or arrowhead design and are made through a molding rather than a wire-bending process. The plastic clips are generally useless for fastening reasonable amounts of paper together, and they tend to bend a few sheets beyond what should be acceptable limits. However, such arrowhead clips continue to cross office desks, and it is reasonable to ask why. Certainly they are nonmagnetic, and this may be one of their selling points. They promise not to threaten any computer data, and perhaps would be gentler on photocopying machines. No doubt the plastic clips could be made economically and in bright colors, but these are definitely not sufficient reasons to use something that simply does not work.

The sadly performing if cheerfully colored plastic invaders of the formerly staid and steely world of paper clips may never get much of a hold on the market, unless some inventor-advocates and manufacturers can remove the serious functional shortcomings and make the plastic and plastic-coated clips work better. They may never have to work as well as a Gem or its relatives, for there may be a cost or appearance advantage to offset a more technically functional disadvantage, but the tradeoffs must be more evenly matched than they are at present if the newest entrants into the realm of paper clips are to survive as functioning artifacts. The competition is very tough, and the Gem has a solid hold on its reputation, if not on its papers.

With paper clips, as with all artifacts, any challenge to the long-established standard will succeed only by calling attention to and overcoming the shortcomings of the Gem. The invention of a new paper clip will not occur in some amorphous dream world devoid of all artifacts save imaginative shapes and styles of bent wire or formed plastic. Rather, any new clip will come out of the crowded past of reality, which is littered with torn and jumbled papers and misshapen challengers to the Gem. Whether a new entrant is a thinner and less expensive Gem look-alike or a restyled Gem "perfected," it will have to be hailed as a "new and improved" paper clip to unseat the acknowledged archetype.

Engineering is invention institutionalized, and engineers engaged in design are inventors who are daily looking for ways to overcome the limitations of what already works, but not quite as well as can be imagined—or is hoped. Whether or not improved designs for computers, bridges, or paper clips come to be patented or incorporated into the technological landscape, they are always explorations of the possible paths along which technology can evolve.

5

Little Things Can Mean a Lot

Many a writer on technology has been struck, in a moment of pause between sentences or an hour of distraction between paragraphs, by the extraordinariness of ordinary things. The push-button telephone, the electronic calculator, the computer on which words such as these are processed are among the more sophisticated things we use, and they can awe into silence those of us who are not electrical engineers. On the other hand, such low-tech objects as pins, thumbtacks, and paper clips are frequently and verbosely praised for their functionality and beauty of line, but are seldom the subject of study, unless it is for the sake of learning how to market something that people use much but consider little. The most common of objects are certainly not generally thought to hold lessons for technological process, prowess, or progress.

But if there are general principles that govern the evolution of technology and artifacts, then the principles must apply equally to the common and to the grand. And how much easier it may be to understand how technology works if we can pursue it in the context of something that is less intimidating than a system that takes teams of engineers years to develop. The individual complexity of supercomputers and skyscrapers, nuclear-power plants and space shuttles, distracts us from the common basic elements of technological development that underlie all things—the great and the small, the seemingly simple and the clearly complex. The individual designer and engineer involved in the creation of large systems is often lost in numerous management shuffles, and the story of the end product is frequently that of a major production with an anonymous if professional cast of thousands, no single one of whom is commonly known to be *the* designer or *the* engineer. But, although the often amateur actors in the little-theater pieces surrounding the design and development of many of our simple objects may also be anonymous as far as myriad consumers are concerned, the plot is usually much easier to follow.

Ironically, the largest and publicly most anonymous engineering structures and systems—like bridges, skyscrapers, airplanes, and power plants—are frequently produced by companies named after people. Thus we have the Burns and Roes, the Brown and Roots, the Bechtels, and the countless regional and local founder-named construction companies that ultimately shape so much of our public space and convey a sense of civic pride and achievement. We have the airplanes called Curtiss-Wrights and McDonnell-Douglases, namesakes of inventors and innovators whose pioneering has, directly or indirectly, given us the space shuttles, superjets, and even corporate jets of today. And we have the Westinghouses and the Edisons that have provided us with the electric-power stations and distribution networks that make modern life so comfortable and convenient. We have the Fords, Chryslers, Mercedes-Benzes, Rolls-Royces, and other automobiles invoking with their grilles-cum-headstones the names of long-gone entrepreneurs. Even the giant corporations called General Electric, General Motors, and General Dynamics can invoke the sense of a leader of manufacturing troops more than evoke the culmination of a concatenation of once individual and private companies.

The names commonly associated with some of the most familiar and cherished products on our desks, on the other hand, are obscure if they are known at all. Items like pins and paper clips certainly do not carry nameplates or medallions memorializing their makers. If we do examine the box that paper clips come in, it tells us our clips were made by Acco or Noesting, which hardly sound like the names of inventors or even people. Many a desk stapler reads Bostitch. Is that someone's name, or what? Modest products tend to have at best a pseudonymity that gives little hint of their inventor, but the brand name of a product and that of the company that makes it often do hold clues as to how the product evolved, and thus can give tremendous insight into the evolution of things. And the stories of their names often parallel those of the products that were developed to solve problems with, if not the downright failure of, pre-existing products.

On a package of Post-it notes, those little yellow slips of paper that stick to everything from irate correspondence to refrigerator doors, we can find the "Scotch" trademark and a bold "3M," which older thingophiles and trivia buffs may recall was once known as the Minnesota Mining and Manufacturing Company. How did such a seemingly self-explanatory and no-nonsense-sounding company get involved with little sticky notepads? Besides, isn't Minnesota peopled by Scandinavians rather than Scots?

In 1902 five businessmen from Two Harbors, Minnesota, formed the Minnesota Mining and Manufacturing Company to quarry what they thought was a local find of corundum, a mineral just short of diamond in hardness and thus a valuable abrasive for grinding-wheel manufacturers. The mineral proved inferior for that application, however, and so in 1905 the fledgling company turned to making sandpaper in Duluth. Difficult years followed, with new financing staving off bankruptcy, but true success in selling sandpaper could come only with a product at least as good as the other guy's.

In 1916 the company's sales manager insisted that a laboratory be formed to carry out experiments and tests to ensure quality control so that salesmen would not be embarrassed by faulty products. The laboratory in time made possible the research and development necessary to produce new and improved items in response to problems experienced by sandpaper users. Whereas a manufacturer's salesmen might say that, after quality control, the raison d'être for a company's laboratory was to respond to customers' needs for new products, engineers might see a laboratory more as a troubleshooting workshop in which to deal with the horror stories of product failures and the general tales of irritating shortcomings brought home by the salesmen. In the course of troubleshooting, new products would naturally evolve to deal with objections to existing products.

In the making of sandpaper, an abrasive material is bonded to a paper backing, and the quality of the product depends not only on the quality of the principal raw materials of grit and paper, but also on how uniformly and securely they can be combined. Hence, to manufacture sandpaper it was necessary to develop an expertise in coating paper with adhesive. Unfortunately, even with good glue, the paper used in early sandpaper fell apart when wet, and so using sandpaper was necessarily a very dry and dusty operation. But in the growing automobile industry, where in the 1920s a considerable amount of sanding was needed to finish the paint on auto bodies, the dust was causing lead poisoning among workers. Making waterproof sandpaper would allow wet sanding, which in turn would cut down on dust, and thus be a great improvement. To remove the failings of existing sandpaper, the Minnesota Mining and Manufacturing Company developed a waterproof paper that one of its young lab technicians, Richard Drew, was asked to take to some St. Paul auto shops, where it might be tested. In doing so, he became aware of another problem.

The new two-tone style of painting automobiles was popular in 1925, but it presented considerable problems for auto manufacturers and body shops alike. In order to get a clean, sharp edge when applying a second paint color, the first had to be masked, of course, and this required that newspaper or butcher paper be fastened to the car body. If shop-brewed glue was used, it would sometimes stick so well that it had to be scraped off, more often than not pulling paint with it. Surgical adhesive tape was sometimes employed, but its cloth backing tended to absorb solvents from the newly sprayed paint and cause the masking materials to stick to the paint they were intended to protect. Clearly, existing means of masking had serious flaws. One day, when he was dropping off a batch of waterproof sandpaper, Drew overheard some body-shop workers cursing two-tone painting. The young technician, who had studied engineering through correspondence-school courses, promised he would make something to solve the problem.

As in the majority of design problems, Drew's objectives were most clearly expressed principally in negative terms: he wished to have a kind of tape whose adhesive would not stick very readily. This not only would allow the tape to be formed in rolls from which it could easily and cleanly be removed, but also should have enabled it to be removed easily from a freshly painted auto body. Stating the problem and finding the right combination of adhesive and paper are two different things, however. The first could have come in a flash at a body shop. The latter took two years of experimenting with oils, resins, and the like, not to mention papers to which they could be applied. After many negative results and suggestions that the problem should be dropped, Drew tried some crepe paper left over from unrelated experiments and found that its crinkled surface proved to be an ideal backing. Samples of the new product were taken by the company's chief chemist to Detroit auto manufacturers, and he returned to Minnesota with orders for three carloads of Drew's masking tape.

According to company lore, the tape came to be called Scotch because on an early batch of two-inch-wide tape the adhesive was applied only to the edges, presumably since this was thought to be sufficient and even perhaps desirable for masking applications. One edge of the tape would hold the paper, the other would adhere to the auto body, and the dry middle would not stick to anything. However, with so little adhesive, the heavy paper pulled the tape off the auto body, and a frustrated painter is said to have told a salesman, "Take this tape back to your stingy Scotch bosses and tell them to put more adhesive on it." Though some company old-timers have labeled the story apocryphal, others give it credence by recalling that the incident "helped spark the inspiration for the name" of the line of pressure-sensitive adhesive tapes that now carry the tartan trademark, presumably not because the manufacturer is stingy with adhesive but, rather, because consumers can use the tape to make economical repairs on so many household items.

Cellophane was another new product in the late 1920s, and its being transparent and waterproof made it ideal for packaging everything from bakery goods to chewing gum. It was even natural to want to package masking tape in cellophane, and so someone in St. Paul was experimenting with it. At the same time, Drew was working on another problem, trying to remove his tape's shortcoming of not being waterproof and thus not being applicable in very moist environments. He got the idea of coating cellophane with his adhesive, which would certainly be a promising new tape to make clear packaging watertight. But sticking an adhesive that works wonderfully on crepe paper onto cellophane is easier said than done, and using existing machinery to manufacture quantities of a new product made of a new material usually involves considerable experimentation and development. In the case of Scotch cellophane tape, Drew's initial attempt to make it waterproof failed to come up to expectations: "It lacked proper balance of adhesiveness, cohesiveness, elasticity and stretchiness. Furthermore, it had to perform in temperatures of 0 to 110 degrees F in humidity of 2 to 95 percent." Not surprisingly, at first it did not, and so presented some well-defined problems to be solved.

After a year of work, Drew did solve the problems, at least to a satisfactory degree for the time, and shiny-backed cellophane tape was *the* transparent tape for many years. It was used for all sorts of mending and attaching jobs, and its yellowing with age, its curling up and coming off with time, and the notorious stubbornness with which it hid its end and tore diagonally off the roll were accepted by users as just the way the tape was—nothing better was available. But inventors and tinkerers like Drew saw each shortcoming as a challenge for improvement, in part because they and their bosses knew that competitors did also. Difficulties in getting Scotch tape off the roll, for example, prompted the development of a dispenser with a built-in serrated edge to cut off a piece squarely and leave a neat edge handy for the next use. (This provides an excellent example of how the need to dispense a product properly and conveniently can give rise to a highly specialized infrastructure.)

As changes in the tape were made, new and improved versions were offered to users who wondered how they had ever gotten along with the old tape. Indeed, the company's own description of the latest version of its product can be read not only as praise for Scotch Magic Transparent Tape but also as an indictment of cellophane tape: "It unwinds easily. You can write on it. You can machine-copy through it. It's water repellent. And, unlike the earlier tape, it won't yellow or ooze adhesive with age." This list of implicit and explicit faults of the "earlier tape" sure makes it sound disgusting and inadequate, but in its day it was the cat's meow. Our expectations for a technology rise with its advancement.

The company that began by making an adequate sandpaper may not have foreseen the nature of its products many years hence, but the accumulation of experience in attaching adhesives of one kind or another to paper and other backings, and a receptivity to new applications of that expertise—and others accumulated along the way—enabled the Minnesota Mining and Manufacturing Company eventually to make tens of thousands of products. Since the old name no longer fully described the diverse output of the giant manufacturer, it came to be known more and more by the abbreviation "3M," and in a recent annual report to the stockholders the full name appears only in the accounting statement.

The characteristic of 3M that enabled it to attain such diversity in its product line is a policy of what has generally come to be called "intrapreneurship." The basic idea is to allow employees of large corporations to behave within the company as they would as individual entrepreneurs in the outside world. A model intrapreneur is Art Fry, a chemical engineer who in 1974 was working in product development at 3M during the week and singing in his church choir on Sundays. He was accustomed to marking the pages in his hymnal with scraps of paper, so that he could quickly locate the songs during the two services at which he sang. The procedure worked fine for the first service, but often by the second one some of the loose scraps of paper had fallen out of their places. Fry, not having noticed this, was sometimes at a loss for words. Now, loose scraps of paper have long been used for bookmarks—some are clearly visible in the foreground of Albrecht Dürer's famous etching of the great humanist Erasmus—and one can safely say that many a bookmark had lost its place in the four and a half centuries between that etching's date of 1526 and the time when Fry reflected on the failure of bookmarks to do all that might be expected of them.

Fry remembered a curious adhesive—a strong and yet easily removed "unglue"—that Spencer Silver, another 3M researcher, had come upon several years earlier in the course of developing very strong and very tacky adhesives. Although it was not suited to solving his immediate problem, Silver felt the unusual adhesive might have some commercial value, and so he demonstrated it to various colleagues, including Fry. At the time, no one had come up with a use for it, and so the formula for the weak adhesive was filed away—until the Monday morning when Fry came to work with the idea of making sticky bookmarks that could also be removed without damaging the book. His initial attempts left some adhesive on the pages, and Fry has surmised that "some of the hymnal pages I tested my first notes on are probably still stuck together." But since it is 3M's policy (and that of other enlightened companies) to allow its engineers to spend a certain percentage of their work time on projects of their own choosing, a practice known as "bootlegging," Fry was able to gain access to the necessary machinery and materials and to spend nearly a year and a half experimenting and refining his idea for sticky—but not-too-sticky—slips of paper that could be used for "temporarily permanent" bookmarks and notes. While Fry wanted bookmarks to stick gently to his pages, he did not want their projecting ends to stick to each other, and so adhesive was applied at one end only. This also served well for repositionable memos and removable notes: with adhesive all over their backs, these would have been as hard to peek under and remove as labels.

Albrecht Dürer's *Portrait of Erasmus,* prominently dated 1526, shows that the great Dutch Renaissance humanist employed scraps of paper as bookmarks. Though they were sure to mark his place as long as the book was latched shut, some bookmarks may have slipped between pages or fallen out as the book was used. About 450 years passed before the use of loose scraps of paper frustrated someone enough to motivate him to persist in inventing a more tenacious paper bookmark, which in turn led to the now familiar Post-it notes.

When Fry thought the stick-and-remove notes were ready, he took samples to the company's marketing people, "who had to accept the idea as being commercially viable and meeting a market need" before any substantial amount of the company's own time or money was to be invested in the product. There was a general lack of enthusiasm for something that would have to sell at a premium price compared with the scratch paper it was intended to replace. (Its removable-note function was believed to hold greater commercial potential than its sticky-bookmark function.) Fry was committed to his brainchild, however, and he finally convinced an office-supply division of 3M to test-market the product, which "met an unperceived need." Early results were not at all promising, but in those cases where samples were distributed, customers became hooked. Though no prior need for the little sticky notes had been articulated, once they were in the hands of office workers all sorts of uses were found, and suddenly people couldn't do without the things. Post-it notes were generally available by mid-1980 and are now ubiquitous, even coming in long, narrow styles to accommodate the vertical writing of Japanese. It might be argued that they have reduced the recycling of scrap paper as scratch paper and bookmarks, but the removable notes do have the conservatory potential of eliminating the use of unsightly and damaging tape and staples for posting notes and announcements in public places.

Years ago, when I would meet my dean walking across campus to the Engineering School, as we approached the building he would invariably remove the numerous announcements of meetings, parties, and kittens for adoption that had been taped or tacked to the door since his last entrance. He carefully peeled off the tape that made posting notices so easy but maintaining an attractive entrance to a school so difficult. The dean explained on more than one occasion how the tape could become difficult to remove if it stayed up over several days and nights, and how it had ruined some freshly painted walls, which had had to be patched and repainted. The dean was not opposed to notices, but to the damage that their attachment did to the main entrance to his school. How he might have loved Post-it notes and dreamed of them in poster sizes.

Post-it notes provide but one example of a technological artifact that has evolved from a perceived failure of existing artifacts to function without frustrating. Again, it is not that form follows function but, rather, that the form of one thing follows from the failure of another thing to function as we would like. Whether it be bookmarks that fail to stay in place or taped-on notes that fail to leave a once-nice surface clean and intact, their failure and perceived failure is what leads to the true evolution of artifacts. That the perception of failure may take centuries to develop, as in the case of loose bookmarks, does not reduce the importance of the principle in shaping our world.

Scrolls were once the standard medium for recording and preserving the written matter of everything from politics to scholarship, and a single scroll was called in Latin a *volumen,* from the verb "to roll." The length of such a volume was limited literally by how long a piece of papyrus could be rolled up or onto the rods that marked its ends. Papyrus was made by laying crosswise slices of the pith of the papyrus plant and pounding or pressing them together until they formed sheets, which in turn could be pasted together end to end to form the long and narrow continuous sheets needed for scrolls. The papyrus sheets were rolled rather than folded up, because the material cracked easily and thus could not for practical purposes be folded over on itself.

If the written word were still preserved on scrolls, going from one end of a long manuscript to another could involve considerable rolling and unrolling. One way of removing this inconvenience of scrolls, while at the same time eliminating the need to form sheets of writing material into long strips, is to form or fold the sheets into leaves of uniform size that can be bound together in some way along one edge. Parchment and vellum, which were formed from the skin of newborn lambs, kids, and calves, could be folded without cracking, and so volumes no longer had to be stored on rolls. With the introduction of paper and the printing press, books multiplied and their binding came to be done more and more efficiently with needle and thread in the folds.

The needle is among the oldest of artifacts, and its usefulness is without question. Yet in certain applications it has severe shortcomings. The infrastructural thimble answers to the problem of pushing a needle through a tough piece of material before the needle itself pierces the finger. And the clever diamond-shaped loop of fine spring wire is a godsend to those of us whose squinting eye can never seem to focus on the more squinting one we are trying to thread. But the needle has also led to the development of many other twentieth-century artifacts that we may hardly recognize as related.

Needles can be thought of as headless pins with a single eye meant to pass readily through anything from the hem of a veil to the hide of a camel, leaving behind only a thread of a clue to their presence. When fully sewn into a garment, a piece of thread can be thought of as a continuous and flexible ghost of a needle, but with no bulging eye to find our most sensitive spot and no point to pierce our skin. Needle and thread not only fashioned the clothes of our ancestors but also gathered printed leaves of paper into signatures and these into volumes; though in this latter application the thread may have been invisible to the reader, it certainly left its mark on the book.

The classic shape of a book's spine derived from the fact that the folds of paper that formed it were thickened by the passes of thread that they contained. To keep a bound book from having a spine much thicker than its other edges, and thus from having the undesirable shape of a wedge that would make stacking and shelving books much less convenient than it is, the sewn spine was rounded and fanned out before binding, so that the threads did not sit directly on top of one another. The boards that formed the hard front and back of the book added enough thickness to rise above the thickest part of the spine, and the hinge of cloth that connected them followed the rounded shape of its contents. The characteristic shapes of books were clearly captured by Dürer in his *Portrait of Erasmus,* with the front edge of their pages fitting the curve of their spines because the paper was trimmed before the spine was formed.

Though today's books may appear to retain the curved look in their spines, it is really just a rounding of the stiffened binding cloth. The book proper has a squarish shape, with a flat front edge as well as a flat spine. This change in form came about because the traditional procedure of sewing books in signatures was time-consuming and costly, and thus failed to be as economical as alternative procedures. The typical book is now "perfect-bound," which means that its sheets are folded in signatures as before but not sewn. Rather, the signatures are gathered and stacked, and trimmed all around to a boxlike shape. Containing no thread in its folds, the stack of paper does not bulge at the spine, and so does not have to be rounded. Instead, it is ground to a rough finish, the better to receive an adhesive similar to the stuff that holds pads of paper together. This procedure was first used in binding cheap paperbacks and has now been almost universally adapted to even the most expensive hardcovers, to the dismay of many an author, reader, and bibliophile. In spite of its name, perfect binding has great failings, not the least of which is that a book so bound is often badly misshapen after a single reading. The modern bookshelf is thus characterized not by a neat ripple of round-ended volumes but by a jagged surface of creased spines. When seen on end, once-read perfect-bound books are sadly skewed reminders of how form follows fortune. Even if this may be to the myopic delight of manufacturers, it can certainly be to the dismay of those who have a sense of form.

In the late nineteenth century, magazines came to be bound by what amounted to sewing with a piece of wire, which could serve as both needle and thread, and a single-wire stitch was certainly much stronger and self-contained than one of cotton. Furthermore, a short piece of bent wire could pierce and hold together more separate sheets of paper, and eventually small booklets and magazines could be made in a single signature with a single stitching operation, known as "saddle stitching." Toward the end of the century, wire stitchers were common in the printing-and-bookbinding industry. Although they were cumbersome machines and took some time to adjust for different thicknesses of work, this was not an unacceptable disadvantage in producing large printings. But for smaller jobs the setup time could be prohibitive, and so a stitching machine that could be adjusted with a slight turn of a screw would lower considerably the cost of printing small runs of small booklets.

Such a machine was built in 1896 by Thomas Briggs, an inventor who lived in the Boston suburb of Arlington. He called his company the Boston Wire Stitcher Company, after the machine it manufactured, and the firm rapidly outgrew its two early homes. In 1904 it settled in a large new factory in East Greenwich, Rhode Island, where the company's descendant flourishes today. Briggs's original machine worked on a conventional principle, which was to feed wire from a head parallel to the seam being stitched, cut off the proper length, bend it into a U-shape, and then drive it into the work and clasp it into what was called a stitch. Because of the size of the feeding head, stitches could not be made closer than twelve inches apart in a single operation. This meant that to bind a small pamphlet took at least two separate stitching operations. In East Greenwich, Briggs developed a machine that fed the wire perpendicular to the seam, cut off a piece, and then turned it before bending it and stitching it into the work. This meant that stitches could be made as close as two inches apart in a single operation, and so binding could proceed at least twice as fast as before.

What made wire-stitching machines so complicated and hence expensive was the mechanism to cut off, turn, and bend the short pieces of wire. To overcome this disadvantage, machines were developed that used individual pieces of wire preformed into a shape that could be driven directly into material being stitched together. The individual pieces were called staples, after the U-shaped pieces of wire with sharp ends that were driven into wooden doors, walls, and posts to secure hooks, hasps, wire, and the like. Although rudimentary stapling machines date from as early as 1877, the first ones had to be fed by hand a staple at a time, and thus were very slow-operating indeed. In 1894 a stapler was introduced that employed a supply chamber into which a line of loose staples could be loaded, but it was a very delicate procedure, for the loose staples had to be pushed off a wooden core onto which they were packed, an operation that had to be done slowly and carefully lest jamming occur. These shortcomings were removed by wrapping a supply of staples in paper around a tin core, thus holding them in place until used; the stapling machine could cut a fresh staple out each time the line was advanced. The driving-and-clinching operation itself was relatively simple and straightforward, principally requiring brute force to push staples through the work and turn them on a sturdy anvil on the back side. Thus, stapling machines could be made inexpensive enough for the smallest printshops and binderies to buy and use, and these were indeed the earliest markets for the new devices.

The first of Briggs's pamphlet-and-magazine staplers were large, freestanding, and foot-operated. They certainly would have been overkill for fastening just a few papers together in an office, and so simple straight pins or the newer wire paper clips continued to be used for that. Hence the Boston Wire Stitcher Company saw business offices as a ready market for a light-duty stapler, and in 1914 offered a desk model priced accordingly. However, the first desk staplers employed loose or paper-wrapped staples, were relatively complicated in their construction, and were prone to jamming. It was not until 1923, at the height of office-efficiency movements, that a simplified desk stapler was introduced and "the use of staples for attaching related papers received its first big push." Soon the company introduced staples glued together in a strip, "which eliminated the disadvantages of handling, loading and feeding which had plagued users of loose staples," and this unpatented idea spread quickly among the growing competition. As stapling machines grew in importance for the Boston Wire Stitcher Company, which had long since moved out of Boston, a distinctive trade name was sought. From the already shorthand name of Boston Stitcher came the contraction Bostitch (pronounced "Boss-titch"), which was registered as a trademark for the stapler line. This name became so prominent that in 1948 the company's name was changed to Bostitch, Inc.

By the early 1930s, desk staplers were smooth-operating little machines indeed, and changes were generally restricted to cosmetic streamlining in keeping with the times. But the new models also incorporated an easier method of loading and could be used as a tacker as well. Thus, the light desk staple, which had the origins of its name, at least, in the U-shaped, double-pointed tacks that for so long had attached hooks to doors and barbed wire to fence posts, was being employed (not always to the benefit of the surfaces so attacked) to fasten signs and notices to bulletin boards, telephone poles, and school walls and doors. This was but one of the hundreds of variations of fasteners made by just one company, whose house history confirms that "new models are always under development, sometimes to do a job that has not been done before, sometimes to do better or faster a job that is already being done." It is especially out of such comparatives that variations in the form of staplers and all technological artifacts evolve.

6

Stick Before Zip

On many a cold winter day I have been frustrated in trying to keep my long woolen scarf wrapped securely around my neck. It seems that the scarf somehow works its way loose over my shoulder with the rhythm of my step, made all the more rapid in the biting wind, and I find myself constantly having to throw the free end back around my shoulder. I have experimented with letting different lengths of scarf hang in front of and behind my body, but I have yet to come upon a foolproof combination of number of wraps around my neck and arrangement of hanging ends that works under all conditions. On the worst of days, I have securely knotted the scarf so that I might avoid the frustration of having constantly to readjust it in the extreme cold. The failure of the scarf to stay in place has driven all my experimentation.

During my winter walks, it is not hard for me to imagine our earliest ancestors experiencing similar frustrations with the animal skins they used to protect themselves against the elements. The skins could be held closed with the hands and arms, of course, but that would have been as inconvenient as my having to hold my scarf in place by keeping a hand on my shoulder. Though this would still allow me to carry a bookbag with my free hand, our earliest ancestors might have preferred to have both hands unencumbered, in order to be ready hunters or to be able to flee more effectively from angered huntees. Since the bulkiness of animal skins would not allow them to be easily tied around the body, alternative means of closure were developed.

Fish bones, pointed fragments of wood, animal bone, or horn would have been found naturally wherever ancient peoples were. But the first threading of a sharp object through two overlapping pieces of skin or fur was an act of invention. The identity of that ancient genius who was so inspired is lost to history, but, somehow, somewhere, garments came to be fastened with pointed pins of bone and horn, which naturally evolved into metal devices.

One great disadvantage of closing garments with straight pins, whatever the material, was that they might be dropped and lost in the course of dressing and undressing or they might work loose during the act of walking or running. Furthermore, the constant insertion and removal of pins would have steadily enlarged holes in garments and hastened their disintegration. This undesirable feature of pins would not have been ameliorated when they came to be used with the woven fabrics that replaced animal skins. Thus, fasteners that were not only less easily dropped and lost but also less likely to wear holes in clothing would have been welcome innovations. Pins might have been tied to garments so the pins were not lost so easily, but this would not have solved the problem that a pin repeatedly stuck into the same place soon wears out its welcome. Such alternatives as the frog-and-loop fastener, whose form might be imagined to have evolved from the pin in as mysterious a way as the form of a natural frog evolves from a tadpole, provided both the advantages of an attached pin and a fixed site of tougher material through which the pin could be repeatedly inserted and removed to close and open the garment.

In ancient times, metal brooches and buckles also came to be used for fastening clothing. While these were fully separable from the garment, they were larger than straight pins and so less likely to be lost, and brooches and buckles held firmly enough not to come loose easily in the course of a day of movement. As long as twenty-five hundred years ago, the Romans developed safety pins, but they seem to have been rediscovered in the mid-nineteenth century. In 1842 Thomas Woodward, of Brooklyn, New York, patented a "manner of constructing shielded pins for securing shawls, diapers, &c.," which he called "the Victorian shielded shawl and diaper pin." His device consisted of a pin hinged to a cupped piece of metal that covers the point, and it is highly suggestive of a modern safety pin. According to Woodward's patent, his shielded pin had distinct advantages over unshielded pins: "It will not become loosened by the motion of the wearer and . . . the point of the pin cannot, by any accident, be caused to puncture, or scratch, the person." However, this pin had no integral spring, and so it had to rely upon the bulk of material compressed within it to hold the point in the shield.

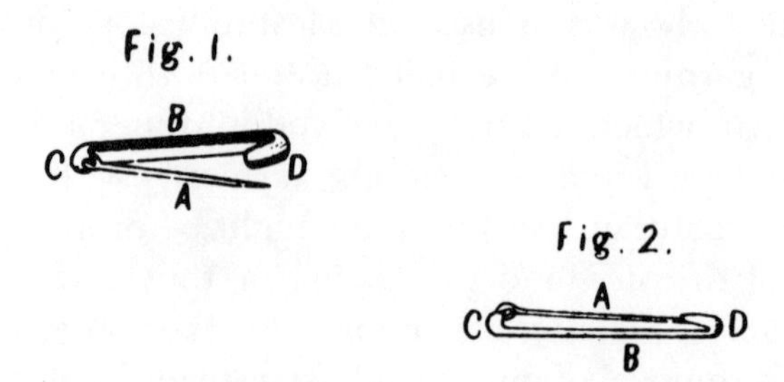

Straight pins of one kind or another were used to fasten garments at least as early as the Bronze Age, and among the age-old problems with the use of such pins was their tendency to work loose and prick the wearer. In 1842 Thomas Woodward patented this "shawl pin," which could be kept in place by having its point pressed against a guard by the bulk of the material; the guard also served to keep the point from sticking the user.

That shortcoming was removed by the "dress-pin" invented by Walter Hunt of New York City and patented in 1849. His pin had "the distinguishing features of . . . one piece of wire or metal combining a spring, and clasp or catch, in which catch, the point of said pin is forced and by its own spring securely retained." Of all the microfilmed patents I have looked at, Hunt's for the safety pin is the only one whose page of illustrations looks like the fragment of a long-lost manuscript. The chipped and broken brittle edges of the original paper document from which the copy was made suggest the curiosity of countless patent examiners, searchers, and inventors to understand the secret of a million-dollar idea. The illustration is famous in part because of the story behind the invention of the new safety pin.

Hunt was a prolific inventor, who was responsible for creating a forerunner of the repeating rifle and the sewing machine. In fact, he actually built the first sewing machine in America, but never patented it because he thought it would destroy jobs. However, Hunt did patent many other items, and for his applications he would naturally have needed to have drawings made. Evidently Hunt was in debt to his draftsman when he proposed that the debt would be forgiven and Hunt would be paid $400 if he assigned to the draftsman the rights to whatever devices Hunt could invent out of an old

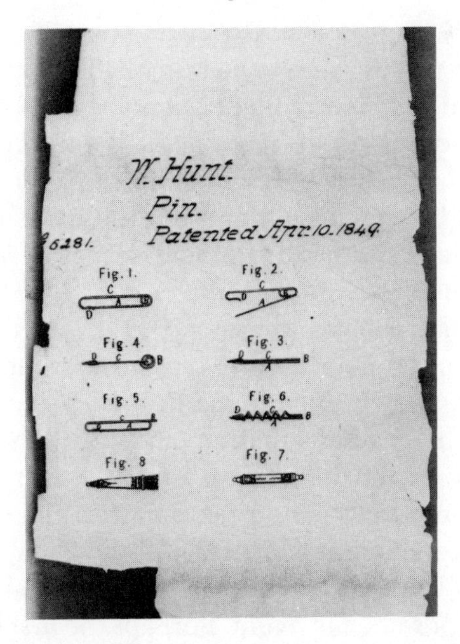

Among the shortcomings of "shawl pins" with no spring of their own was the need to gather just the right bulk of material so that the pin would be kept firmly latched and yet not be bent out of shape. Although self-springing safety pins appear to have existed in Roman times, in 1849 Walter Hunt received a patent for a modern version. His patent, which illustrates various decorative embodiments of the basic idea, is considered a well-thumbed classic among inventors, as attested to by the condition of the copy from which this microfilmed image was made.

piece of wire. The safety pin was the product of three hours of twisting.

Although the drawings of Hunt's patent are not signed, we can assume that the illustrator was one of the assignees, "Wm. Richardson" or "Jno. Richardson." But, whoever profited from the patent, the inventor clearly believed he had overcome the failings of preexisting fasteners, for he declared the safety pin to be "more secure and durable than any other plan of a clasp pin, heretofore in use, there being no joint to break or pivot to wear or get loose as in other plans." Furthermore, the self-sprung pin could be used "without danger of bending . . . or wounding the fingers." It clearly eliminated a lot of the failings of earlier devices.

Whether Sumerian or of later design, the safety pin and other loose and separate fasteners were not practical for closing very tight-fitting garments such as came to be fashionable in the Middle Ages. Body-hugging garments were made possible by the development of fasteners like hooks and eyes and laces. Hooks and eyes had the advantage that they could be fastened quickly, but they were relatively bulky and liable to snag things on the hook. Lacings, on the other hand, while not as bulky and not subject to snagging, took a relatively long time to close up tight.

Buttons and buttonholes were one kind of compromise that removed many of the objections to earlier fasteners. Although the button had been known since Roman times, being inserted into a loop sewn onto the edge of a mating piece of garment, the buttonhole as we know it did not evolve until the thirteenth century, perhaps in response to the failure of a button and loop to make as tight a closure as one might like on a cold windy day, or in response to the fragility of loops and their propensity to break when one was getting dressed for some big event. Perhaps the first buttonhole was actually hastily improvised with a knife or scissors in response to a loop's breaking at the wrong time. But the unreinforced buttonhole would have torn open wider and wider with use, and thus would eventually have failed to hold its button very securely. This shortcoming might easily have led to the reinforcing provided by the now familiar specialized buttonhole stitching.

Even with the advantage that they were less likely than hooks to snag, buttons definitely still took some time to be mated with buttonholes. Nevertheless, an abundance of buttons on garments became a sign of fashion in fourteenth- and fifteenth-century Europe; and the contrasting dressing customs of privileged men and privileged women of that time are generally believed to be responsible for the fact that even today men's clothes button differently from women's. Since, presumptively, most people have always preferred the right hand, a man dressing himself would naturally have favored his right hand for manipulating a button through a buttonhole. Hence, the buttons on men's garments, even if at first attached randomly to one side or the other, would soon have migrated to the man's right-hand side. The most fashionable women, however, were commonly dressed by maids, who naturally faced their mistresses while hooking or buttoning them up. Therefore, buttons would have migrated to the side of a lady's garment that corresponded to the facing maid's right. Any other arrangement would have been inefficient.

Whatever the origins of their orientation, buttons on garments could generally each be fastened relatively quickly. Yet there tended to be so many of them, for a tight closure could not be achieved unless the buttons were very closely spaced, and this was especially important for shoes. But the fingers were not a very effective tool for coaxing the crowded buttons through small buttonholes, and so the buttonhook—a crooked little metal finger—was developed to reach through the buttonhole, grasp the button, and pull it through. With practice, this action could be done quickly and thus had distinct advantages over laces. (Snap fasteners were invented in the nineteenth century and provided the further advantage of being quicker to close and open, and without a special tool, but they were not so strong as button or lace fastenings, and hence not so suitable for shoes, and they tended to wear out faster with repeated use.)

High-button shoes were not only fashionable in the nineteenth century, they were very practical for walking about unpaved streets that were alternately dusty and muddy and always littered with the droppings of horses. But chief among their shortcomings must have been the time it took to button shoes up, for, no matter how deft one could become with a buttonhook, it would take time and attention to insert the hook in each of the twenty-odd holes, hook each of the twenty-odd buttons, pull each button through its buttonhole, and release it with the proper twist of the hook before continuing on. And that would only have fastened one shoe. Although the design of the very specialized hook itself shows little variation, the many handle designs attest that this indispensable utilitarian tool was soon as common and yet as individualized an object on a lady's dresser as was the dinner fork on her table. If buttons should come undone during the day, one might have to have a buttonhook at the ready, and so designs for the purse also multiplied. Yet, because buttoning shoes was something done every day by everyone, including those with inventive turns of mind and dreams of striking it rich, any disadvantage of the process provided a problem to attack.

Though not needed in any absolute sense, the ultimate shoe fastener might be imagined to be one that closed and opened in a single action that took as little time as possible and even less attention. Just as such a device would be invented in response to the shortcomings of the existing scheme for fastening shoes, so the shortcomings of successive stages of the invention itself would drive its perfection. But the process would take decades and not a little money and patience from financial backers.

Something recognizable as a zipper was patented in 1851 by Elias Howe, the inventor of the sewing machine. But Howe's "automatic continuous clothing closure," which consisted of "a series of clasps united by a connecting cord running or sliding upon ribs," was never marketed, and the idea seems to have been forgotten for almost half a century. Indeed, the zipper as we know it today did not begin a fruitful line of evolution until the last decade of the nineteenth century, even though inventors and people from all walks of life were daily reminded of the frustrations of fastening high-button shoes. Once they were put on and buttoned, the feet were imprisoned for the day, unless one wanted to redo several dozen buttons, a task to be avoided even with the assistance of a buttonhook. There appeared to be no more hope of speeding up the buttoning process than there was of accelerating the insertion of a line of hooks in their respective eyes, for the motion of inserting individual hooks into mating eyes, or buttons into mating buttonholes, was across the gap to be closed, whereas the motion of the hand in progressing to close the gap was along it.

The idea of arranging a chain of clasps that could be opened and closed automatically with the single motion of a movable guide that could be slid along the seam was the brainstorm of a mechanical engineer from the Midwest who at one time was being granted patents at the rate of two per year, many for such things as engines and transmissions. In addition to improving the speed and efficiency of the young automobile, Whitcomb Judson was hopeful that with his slider device a simple rapid motion could close and unclose the seams in high-topped shoes. A drawing on one of Judson's first patents for an early zipper actually reveals wire hooks that are pulled together and locked across the seam by an advancing slider. A second patent, issued the same day in 1893, shows a variation on the fastening device, embodying "certain new and useful Improvements in Clasp Lockers and Unlockers for Shoes, &c." The word "zipper" would not be used to describe such a device for three decades.

But even the snappiest of names would not have been sufficient to bring the invention to fruition, for in the 1890s, as in the 1990s, an engineer with a patent was nothing more than that, if there was no capital to develop and market the idea. Fortunately, Judson had

(No Model.)

W. L. JUDSON.

CLASP LOCKER OR UNLOCKER FOR SHOES.

No. 504,038. **Patented Aug. 29, 1893.**

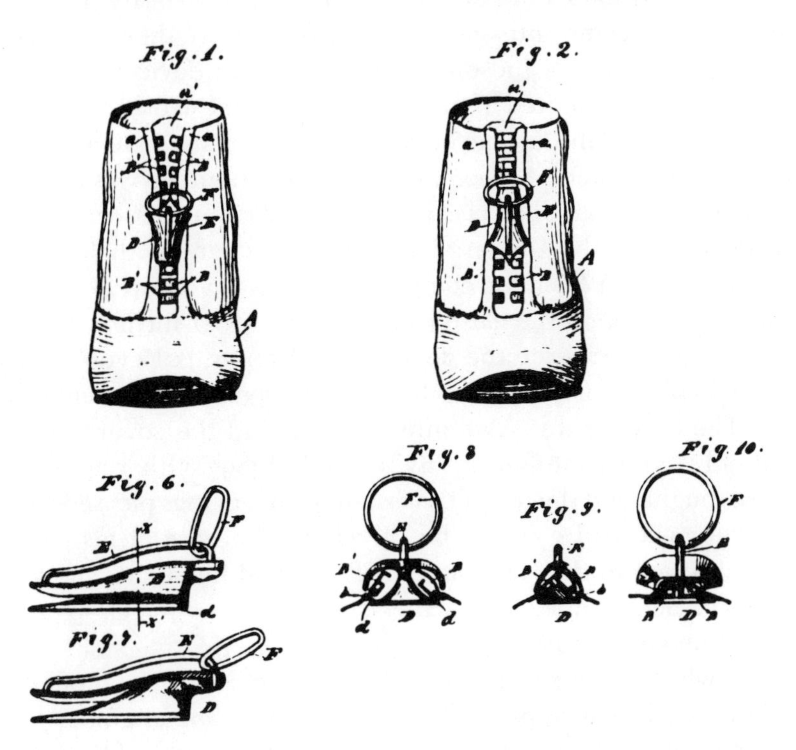

The first patent for a "clasp locker or unlocker for shoes" was issued to Whitcomb Judson of Chicago in 1893 and was illustrated with a buttonless shoe. The invention was motivated by the long-familiar complaint that high-buttoned shoes took so much time to put on and take off.

earlier met Lewis Walker, a young lawyer from western Pennsylvania, who had become interested in the inventor's idea for a street railway system powered by compressed air. The scheme seemed promising for the transportation of oil and coal, which Walker knew would interest his Pennsylvania banker brother-in-law, whose family had benefited greatly from the 1859 discovery of oil in their own backyard. The welcome capital enabled Judson to set up experimental versions of his compression railway in Washington, D.C., and New York City, but the growing application of electric power soon pushed aside schemes like Judson's, and the growing failure of businesses left Walker's in-laws in financial disarray. In the meantime, Walker had come into some money from his father and became interested in one of Judson's more pedestrian devices.

Judson had exhibited his new invention at the World's Columbian Exposition held in Chicago in 1893 and wore on his boots a prototype of the clasp lockers. As soon as Walker saw them, he was convinced of their promise. He had clasp lockers installed on his own boots, and in 1894 set up the Universal Fastener Company with Judson and another partner from the compression-railway venture. Judson continued to work on the clasp locker and took out further patents in 1896, now referring to the device as a fastener, but even the latest model looked bulky, and this limited its appeal to shoe manufacturers. The fastener was sewn into mailbags, but the government had put only twenty into service by the end of 1897. Other applications were sought, and the use of the fasteners in leggings pleased Colonel Walker, who had come to be called that because of his longtime commitment to the National Guard. He had for some time felt his uniform lacked military smartness, and he hoped the clasp fastener would make for a better fit.

But while the financial partners were calculating the profit promised by each new application of the fastener, the engineer Judson was working to "perfect the details," including those of the machinery that would have to mass-produce the fastener if it was to be truly profitable. The application of the fastener to corsets, for example, required it to separate at both ends, and so Judson had to develop a new starting device, since the slider on the original design did not allow the ends to be separated. A saddened Colonel Walker once remarked that "Judson's way of meeting a difficulty was to add invention after invention to his already large supply," but Walker knew as well as Judson that with each new application the shortcomings of the last design were likely to become more evident. Unfortunately, there seemed to be no end to the chain, as the engineer tackled more and more ambitious problems: "Judson's activities were expensive. They tended to create more problems than they solved."

One of Judson's new machines was invented in 1901, and it was to "connect a series of fastening elements—hooks and links—automati-

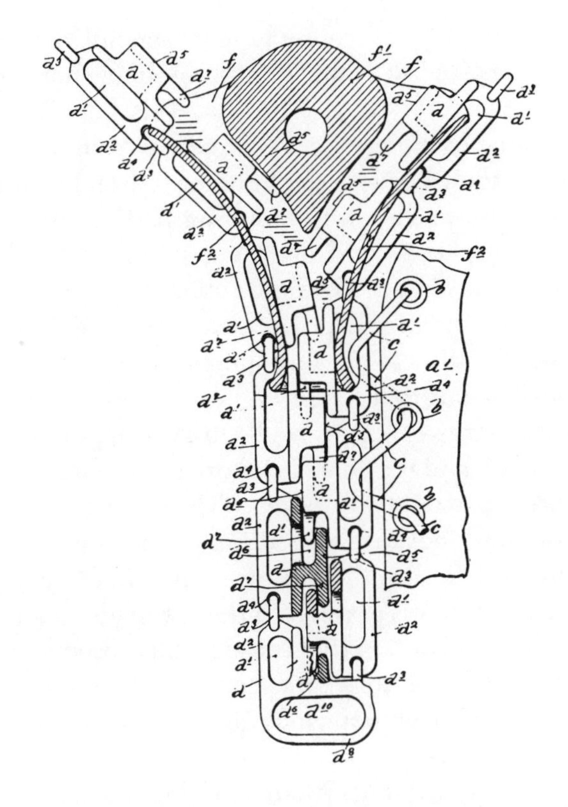

One of a long series of alternate forms of Whitcomb Judson's slide fastener was patented in 1896. Though each version of the fastener appeared to have advantages over earlier ones, frustration in manufacturing and using the devices kept them from really catching on. The difficulty Judson faced in perfecting the slide fastener is suggested by the fact that no one but he was issued a related patent until after 1905.

cally into a kind of chain," but the machine proved too complicated to be useful. With investors disheartened, the Universal Fastener Company became dormant, and even before the new machine would be patented, a new company was formed, the Fastener Manufacturing and Machine Company. Under this company, the fastener continued to be developed, and eventually, "instead of linking the fastening elements directly in a chain, Judson clamped them along the beaded edge of a fabric." This not only removed the complications of prior processes, but, more important, "the finished product could be attached to a garment by a sewing machine. Gone was the tedious necessity of having to sew each link of the old chain fastener to a garment by hand." Thus, another shortcoming was removed.

In 1904 the name of the firm was changed to Automatic Hook and Eye Company, largely because the fastener Judson had finally come up with as suitable for general marketing had fastening elements that resembled old hooks and eyes, but with the hooks pointing along the seam to be closed. The new automatic fastener was called C-curity, to emphasize its advantages over the hooks and eyes that had to be sewn on and closed and opened individually—certainly not automatically—but that too often popped open at inopportune times. Advertisements for the new C-curity fastener celebrated its advantages: "A pull and it's done! No more open skirts . . . Your skirt is always securely and neatly fastened." The device was also termed a "placket fastener," because, according to the company's etymology, "the word placket itself meant woman when it first came into the language. Later it was applied to the slit made in a garment to facilitate putting it on, and in that sense it still is used by the trade."

Unfortunately, for all its names and advertising claims, the C-curity fastener was itself notorious for "popping open at the most inconvenient moments. Worse, when such an accident occurred, the slider became locked in its position at the end of the chain. The only way to get the garment off was to cut it off or cut the fastener out." Furthermore, according to the company's own history, operating the devices wasn't as easy as "a pull and it's done":

> A leaflet printed in March, 1906, tacitly confessed to many difficulties. The instructions for applying the fastener were wordy and complicated. The sponsors of C-curity betrayed their own lack of security by stating: "Customers will confer a favor on us by reporting any difficulty in applying fastener, in which case we will send more detailed instructions." The "instructions for using" were not merely wordy but worried.

As with all artifacts, it was the difficulties encountered in using each successive version of the automatic fastener that led to modifications designed to remove those difficulties. And in this case the modifications had evolved the form back very close to the hook and eye that had generated its departure. Judson had followed a long, circuitous route before he turned his hooks and eyes so that they were oriented along pieces of beaded tape, and, like many an impatient inventor familiar with the habits of his own maturing brainchild, he was able to get it to work pretty well in the laboratory. But customers were not so gentle with the inventor's baby and used it the way it was advertised to be used. And the manufacturer knew they would, because the customers were being asked to help identify problems and difficulties the engineer Judson either could not identify or overlooked in his zeal. But "the hook and eye principle, as applied to the fastener, never ceased to be a complete nuisance," and if the Automatic Hook and Eye Company was ever to succeed, it had to respond to the objections to the device that was central to its name. Either better hooks and eyes would have to be fashioned into automatic fasteners that worked reliably, or the hooks and eyes would have to be replaced by something further evolved mechanically.

The man who was to accomplish what Judson did not was born in Sweden in 1880 and named Otto Frederick Gideon Sundback. His parents owned rich farm and timber lands and so were able to send their technically inclined son to school in Germany, where he received his electrical-engineering degree in 1903. After returning home to fulfill his military service, young Sundback emigrated to the United States, where at the time there were few engineering schools but a growing industrial economy that demanded engineers. Sundback dropped as many of the European trappings of his name as practicable, preferring to be known simply as G. Sundback, and found a job with the Westinghouse Electric Corporation near Pittsburgh, where he worked on designs for the giant turbo-generators for the Niagara Falls power plant.

Pittsburgh was not far from Meadville, Pennsylvania, where most of the financial backers of the Automatic Hook and Eye Company resided, and Sundback's path eventually crossed some of theirs. He was having troubles with his superior at Westinghouse, and so agreed to visit Automatic's factory in Hoboken, New Jersey, for an interview. There he met P. A. Aronson, a highly skilled mechanic whose job it was "to keep Judson's machine running long enough and steadily enough so that its defects could be diagnosed and cured." While in Hoboken, Sundback evidently also met Aronson's daughter, Elvira, whom he later married. Whether for love or the dream of money or the simple fascination and complex challenge of the automatic fastener, in about 1908 Sundback began what was to be a long association with Automatic Hook and Eye and its descendants. The official version is: "His shrewd eye caught some of the defects of manufacture, but the technical expert within him said that the defects could be corrected and the perfectionist within him demanded the opportunity to do so. He took the job."

In Sundback's own words, he became "fully saturated" with the problems of the fastener and he often lay awake half the night "trying to find a way out." He first tackled the C-curity fastener's "trick of popping open" and developed an extension of the eye to enclose the hook completely. Since "C-curity had damaged the company's prestige" so, the improved placket fastener was quickly marketed under the name Plako as "the C-curity Fastener made perfect," even before a patent application was filed. (U.S. Patent No. 1,060,378, issued in 1913, is now often taken as the milestone marking the introduction of the zipper.) Advertising copy proclaimed that "buttons, hooks and clasps are disappearing before Plako," but the company's euphoria proved to be short-lived. Sundback himself told the "rueful story of how the secretary of the company had gone out one evening with a Plako in his trousers and had to hurry home seeking a safety pin." There were many failings still to overcome, and "complaints piled up on the engineer's table."

Though Automatic Hook and Eye owned the patent rights in the United States, the company had agreed that Sundback would retain foreign rights. By 1910 his father-in-law was in Paris and had found backing for a French factory to make "Le Ferme-Tout Américain," but World War I interrupted the enterprise. Things were going badly in America as well, with the passing of the days of steel at five cents per pound and labor at six dollars a week. The staff at Automatic Hook and Eye had been reduced to Sundback and one other person, and Sundback served as executive, engineer, factory superintendent, and office boy. Thus it was he who had to convince a salesman to supply more raw materials when several thousand dollars was already owed to John A. Roebling's Sons, the company that has provided steel wire for everything from failing fasteners to successful suspension bridges. To pay printing bills, Sundback repaired a printer's machines and on one occasion created for him a machine for making paper clips. But there also seemed always to be new backers appearing at the doorstep. James O'Neill, the father of the playwright, was a quick-change artist who toured the country playing in *The Count of Monte Cristo* and thought the Plako fastener was a godsend. O'Neill bought stock in the company and took an interest in developments.

In contrast to such support, Sundback suffered terrible personal setbacks. He was crushed when Elvira died following childbirth, and could go on in his intense grief only by giving his total attention over to the problems with the fastener. He eventually took "a radical departure from all previous forms the device had taken," and focused on eliminating the troublesome hooks that had always proved "fatal" to the design:

> To one side of his new model he now put spring clips or jaws which clamped around a beaded edge of the tape on the other side. The slider was designed to wedge these clips apart, in its upward progress, and force the beaded edge into the opened jaws. The jaws then snapped around the bead. . . . Exit the hook.

Perhaps Sundback got the idea while working on the machine to make paper clips. But, whatever the inspiration, a patent for this new slide fastener was applied for in 1912 and granted in 1917. Colonel Walker was delighted and fascinated with early handmade samples and called the mechanism the "hidden hook." But Sundback was less sanguine in letters to the Colonel, reminding him that finances were so low that the factory had been shut down: "There is hardly any doubt in my mind that the new hidden hook will replace Plako, but before we get ready to fill orders we will want some stock and facilities to manufacture the hidden hook and that is a few months off still." And a few weeks later he wrote:

> It doesn't seem to me that the hidden hook is right for the trade as yet even if the steel and tape were right as to quality. I have found weak points. The catching is quite serious and can only be remedied through some additions to the outside which may give it a rather clumsy appearance.

In spite of the engineer's worries over everything ranging from performance to aesthetics, Colonel Walker still was not discouraged. However, by 1913 some of the early patents were close to running out, and so the financial backers were interested in reorganizing the company. At their annual meeting, the stockholders in Automatic Hook and Eye agreed to sell all the company's assets, and the Hookless Fastener Company came into existence shortly thereafter. Sundback moved the factory from Hoboken to a little barn in Meadville, which paid scant attention to the fact that "an obscure company, engaged in the manufacture of an unfamiliar gadget, had come to town." But many of those who knew of Walker's obsession with hookless fasteners and the like, on seeing him coming down the street, would whisper to their companions, "Cross over quickly. Here comes the Colonel. He'll try to sell us shares in his gadget." In the meantime, Sundback made the best of his cramped quarters and "set about redesigning his machinery, making innumerable experiments."

Sundback had been anxious to get down to work after the move to Meadville, for, although he was disappointed in the spring-clips device, which had come to be known as Hookless No. 1, he had since come up with a new scheme, Hookless No. 2—another "radical departure in principle from the design of earlier slide fasteners." This one he described as being "built up of nested, cup-shaped members." Furthermore, Sundback also accomplished, in the essential machine for making the device, a simplicity to match the fastener's operation. "The interlocking members could be stamped out of metal in one process." When he announced this breakthrough to his backers in December 1913, Sundback confessed that he had never known men to "take anything so calmly," but he also knew that the Colonel especially had never doubted that the fastener concept would someday work, and so, when it did, it was anticlimactic.

The history and operation of the slide fastener was once the cover story for *Scientific American.* While descriptions of the machines that now manufacture the indispensable item by the billions take up much of the article, the cover illustration is a closeup of a fastener of essentially the same design as Sundback's 1913 breakthrough. The slider that usually obscures the operating principle behind the device is removed, and the interlocking teeth are shown in the process of nesting into each other. Each tooth is really shaped somewhat like a deep-bowled spoon, and is more properly called and described as a scoop, for it has been stamped out so that the top is more or less convex and the bottom concave. In the act of closing the fastener, the slider functions as a guide, first pulling the scoops together and then channeling them with just the right orientation so that they alternately nest left into right and right into left as the slider passes over them. When the scoops have all been interlocked, the closing

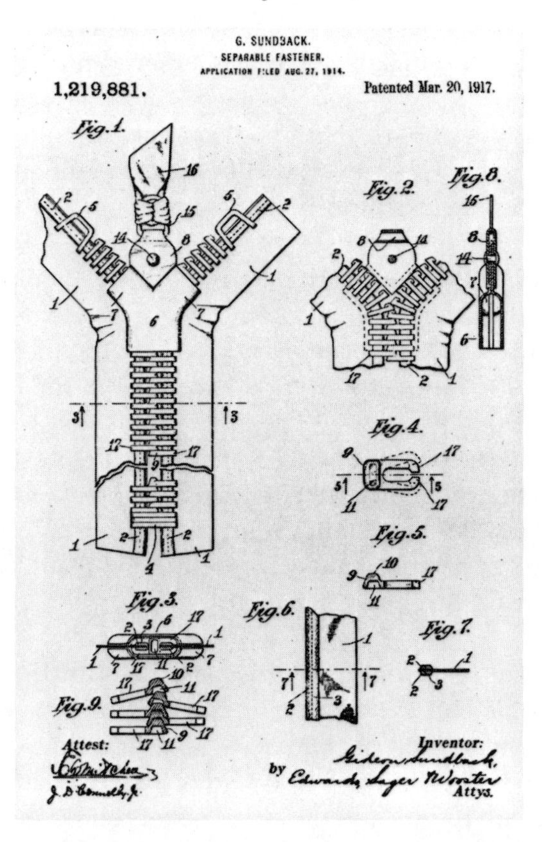

Gideon Sundback was an engineer recruited in 1906 to develop a reliable slide fastener and the machinery to manufacture it. After working for years with various devices that essentially continued to engage hooks in eyes, in 1913 he came up with the idea for a "hookless fastener." The first version was as frustrating to deal with as all the others, but the definitive patent, shown here, was finally awarded in 1917.

is secure (yet flexible) but can easily be opened by pulling the slider in the opposite direction.

Six months after his announcement of a device based on this new principle, Sundback thought he had all the machinery fine-tuned and everything ready to go into full-scale production. The Colonel had planned a party for the big day, but when the power was turned on, "the fool machine wouldn't work." It spat out two inches of fastener and stopped. Eventually it did work, of course, and quantities of Hookless Fastener No. 2 were ready to be sold. But, unlike its predecessor, the Plako, which was "offered merely as a novelty item in a peddler's pack," the "Hookless No. 2 was to be sold directly to manufacturers who were in a position to put it into use on a large scale." Walker knew, however, that he had to point out to the stockholders the problems still to be overcome:

> First, a demand had to be created and then manufacturers of garments and other objects of common use had to be persuaded that the fasteners had become a necessity to them. The demand may be said to have existed for a long time in the unconscious minds of people who were tired of buttons that came off and snaps that wore out and buckles that rattled. But it lay buried under a dead weight of custom and inertia. Manufacturers were positively hostile. They didn't want to face the many challenges of redesign, of drastic changes in methods of manufacture and, most particularly, of additional cost.

The "mother of invention" may have been in the unconscious minds of people, but even if inventors sometimes have an Oedipus complex, manufacturers seldom do. Walker was fully aware of what really was operating in the two decades since he had first encountered Whitcomb Judson and the clasp lockers on his shoes. Certainly there were problems with buttons, snaps, and buckles, but there are problems with everything, including hookless fasteners. Until an inventor could come up with something that did not have the problems of the old and that itself had so much promise and, perhaps, pizzazz that its own problems would be overlooked, "necessity" was an unnecessary word. And Walker also knew that, even when Sundback had broken through all the technical roadblocks, some of the greatest obstacles still lay ahead, square in the face of salesmen who had to approach manufacturers and ask them to spend money redesigning their products, their machinery, and their budgets.

Colonel Walker's two sons, Lewis Walker, Jr., and Wallace Delamater Walker, were enlisted as missionaries, and the latter sold the first four Hookless Fasteners No. 2 on October 28, 1914. The entire proceeds—one dollar—was placed in an envelope and signed by Sundback, thus culminating eight years of engineering development on his part. He was not the only one who had tried to improve on Judson's idea of twenty years earlier, of course. There had in fact been a host of others, women as well as men. Josephine Calhoun, of

Tampa, Florida, received a patent for a variation on the C-curity that she designed in 1907. That same year, Frank Canfield of Denver, Colorado, came up with a system of claws that closed on spherical knobs. And it was not only in America that inventors were busy. Perhaps the idea closest to Sundback's final solution was patented by Katharina Kuhn-Moos and Henri Forster of Zurich in 1912. But none of these ideas turned into manufactured products the way the Hookless No. 2 did.

To a manufacturer, having a warehouse full of fasteners is not as satisfying as having a backlog of orders. And orders for the hookless fastener came very slowly at first. A buyer at McCreery's Department Store in nearby Pittsburgh called the Hookless "the perfect fastener for skirts and suits," which saved time for saleswoman and customer alike in the fitting room; she insisted that the fastener be used by garment manufacturers on all skirts made for McCreery's. But not many buyers followed suit. Few wanted to risk their reputations on Sundback's untried technology: "It worked perfectly in ideal conditions. It could be produced en masse at a reasonable price. But it still had to be tested by continuous use in the hands of amateurs." Sundback understood the reluctance of the marketplace, and he was responsive: When the slider proved too weak, he strengthened it. When a new application was hinted, he adapted the Hookless accordingly. But as late as 1915 the Hookless Fastener Company was faced with the question of "how to create a demand . . . for something which most people had never seen and few had ever dreamed of." To survive, the company was forced to give assurances to manufacturers originating new applications that no competing product would be supplied with hookless fasteners.

The manufacturing technology had been all but perfected, with the Meadville factory able to produce 1,630 fasteners a day, and not a single defective one in the batch. But just as orders were beginning to come in, the war slowed the supply of raw materials, and potential customers began to lose interest in the fastener when orders could not be filled promptly. However, the war proved also to be an opportunity, for money belts closed with hookless fasteners sold to army and navy men as fast as the belts could be produced by the Ewing Manufacturing Company, which by mid-1918 was ordering fifty gross of fasteners at a time. Other wartime applications followed, including flying suits for the air corps, which without buttons could be produced with less material and could be made windproof.

When the suits were tested by the navy, only the fastener passed. It was soon being used in life-preserving vests, and the government was releasing metal to be made into hookless fasteners.

But with the Armistice, demand for fasteners declined. There was no more market for money belts and life vests, and the clothing industry was still not interested. Although the Hookless had proved its functional competitiveness, in order to compete successfully with conventional closing devices the fastener had to become competitive in price. Sundback realized that this could only be done through more efficient manufacturing, and so he developed a process that used specially formed wire and what he called the S-L machine. The letters stand for "scrapless," and the machine in a "continuous operation slices off thin pieces of the Y-shaped wire, puts a pocket on one side of the scoop and a projection on the other, then closes the inner section of each Y-shape around the corded edge of the tape as it is fed into the machine. Result: no waste at all." In fact, fasteners could be made with only 41 percent of the metal used earlier. Among the first products to benefit from the competitively priced fasteners was the Locktite tobacco pouch, which was advertised as the handiest pouch made, with "no strings—no buttons." By the end of 1921, shipments of fasteners to the pouch company were exceeding seven thousand per week. Hookless Fastener Company had long since progressed beyond Hookless No. 2 and had added *Factory* No. 3.

In 1921 the B. F. Goodrich Company of Akron, Ohio, ordered a small number of fasteners. Within a few days of the order's being filled, Goodrich asked if Hookless could deliver 170,000 fasteners "within the near future." Since this exceeded the entire output for the previous year, the query was a "stunner," to say the least, and the order was not filled. The capacity of the factory was already spoken for and, in any case, Goodrich was unwilling to reveal what they wanted the fasteners for. Goodrich continued to ask for small quantities, which it finally confessed it was installing in rubber galoshes in which its clerks were walking around the office—in the heat of summer—to test the durability of the fastener. In the winter of 1922, Goodrich salesmen tested the fasteners in galoshes worn under more inclement conditions. As shortcomings were discovered, they were reported to Hookless engineers for correction. The new Goodrich product was finally announced: "The Mystik Boot with the patented Hookless Fastener. Opens with a pull. Closes with a pull." But salesmen did not like the name of the boot. It did not suggest a very practical article.

"Zip" is a word that came into use in the latter part of the nineteenth century to suggest "a light sharp sound such as that produced by a bullet or other small or slender object passing rapidly through the air or through some obstacle . . . or a movement accompanied by such sound." According to one account, when Goodrich salesmen complained at a sales conference about the fanciful nature of the brand name "Mystik," the company president, Bertram G. Work, himself familiar with the sound made when he closed up the new galoshes, is reported to have replied, "What we need is an action word . . . something that will dramatize the way the thing zips." And then he quickly added, "Why not call it the *zipper*?" And so in 1923 Zippers were first advertised as being "made only by Goodrich," which registered the word as a trademark for the Zipper Boot. But usage respects no company's rights, and eventually "zipper" came to be the common generic name for the device that is more properly called a "slide fastener."

That winter, Goodrich sold almost half a million Zippers, and throughout the mid-1920s it agreed to buy a minimum of a million fasteners per year from Hookless. The name "hookless fastener" seemed archaic beside "zipper," and, showing no sentiment for the fact that it was attention to faults that had brought success, the Hookless Company now felt that its name had a "negative implication." Since "zipper" was Goodrich's, a new trademark was sought, one that emphasized "positive qualities." After rejecting the likes of "Utilok" and "Bobolink," Hookless selected "Talon" as the name for its slide fastener: "Everything about it seemed right. The elements of the fastener were surely like the claws of the eagle, gripping with firmness." In 1937 the trademark would be adopted for the company name.

By 1930 twenty million Talons per year were being sold for everything from pencil cases to motorboat engine covers, but remained virtually unused in women's dresses or men's trousers. The apparel industry would remain conservative until the mid- to late 1930s. Elsa Schiaparelli was among the first clothing designers to use the slide fastener in a big way, and her 1935 spring collection was described by *The New Yorker* as "dripping with zippers." Soon after, an aggressive and successful advertising campaign promoted zippers through humor, including that of James Thurber, and inciting fear—

The B. F. Goodrich Company introduced the Mystik Boot "with the patented Hookless Fastener" for the winter-1923 season, but salesmen did not care for the name Mystik. Goodrich's president agreed that what was needed was "an action word [to] dramatize the way the thing zips" and came up with "zipper." The word was registered as a trademark by the rubber company for its "Zipper Boot" but soon came to be used as the common name for the slide fastener itself.

of the embarrassment of "gap-osis" caused by skin and undergarments showing between snaps and buttons. The widespread adoption of zippers into clothing assured the future of Talon, Inc., as well as that of the growing competition.

If form follows function, then it follows a very circuitous and costly route, as the development of the slide fastener illustrates. The function of today's zipper was as evident to Elias Howe in the mid-nineteenth century as it was to the host of later inventors who were also seeking a workable "automatic continuous clothing closure." But the form to realize that function was far from self-evident, as shown by Judson's clasp locker, Sundback's hookless fastener and scoop-toothed slide fastener, and, most recently, zippers that in place of metal teeth have plastic spirals and shapes that hardly resemble clasps, hooks, or scoops. And there is no telling what today's zippers might look like if one of the many men and women who patented the beginnings of another form of zipper had lain awake as many nights as Sundback did—thinking about how the operating problems with their cursed thing could be removed by a further improvement—or if they had had the benefit of an angel with the financial stamina of Colonel Walker. But, with or without such benefits, like the form of many a now familiar artifact, that of what has come to be known as the zipper certainly did not follow directly from function. The form clearly followed from the correction of failure after failure.

7

Tools Make Tools

Few classes of artifacts exhibit as much diversity and specialization of form as the tools of the crafts and trades. Perhaps this should come as little surprise, for tools are generally acknowledged to be the first artifacts of civilization, and hence they have had the longest time to evolve. Furthermore, because of their very nature, as the artifacts with which are fashioned all other artifacts, tools have a special place in the world of things made.

Through the ages, the professional users of tools by and large have not needed to, been able to, or wanted to talk to outsiders about their implements. They did not need to because tools themselves are used to make other tools, and thus users could very often fashion a new tool with their old ones. If they did need to communicate the design for a new tool to someone outside their trade, such as a blacksmith, they could do so without having to reveal the tool's intended use. Moreover, users in the past were often illiterate and hence ill prepared to describe where and how ideas for new tools originated. Besides, the inventive process of conceiving a new tool was often a nonverbal one. Finally, craftsmen were unwilling to share information about their specialized tools because to do so would have been to give up their competitive edge and their value to those outside the craft.

A telling anecdote about the craftsman's mind and the evolution of tools of the trade has been related by George Sturt in his memoir of the nineteenth-century English farmer and potter William Smith. Although the objects that craftsmen sit on while working might not commonly be thought of as tools, their design can affect the efficiency and smoothness of working as surely as do knives and hammers. Sturt found it "odd" that some of Smith's potting furniture had been given names, and so his attention was drawn to them in the course of describing the use of some stools:

> One stool was called "Broad-ass." Sometimes the potter himself, not finding this stool in his workshop, would sing out, "Bring me Broad-ass." Another stool went by the name of "Old Cockety." But perhaps the most useful of the three, and not the least quaintly named, was a one-legged stool known as "Nobody."
>
> "Nobody" was invented, or introduced at the Farnborough pottery, by one Ninety Harris. . . . It was when the workshops were being extended . . . that Ninety, a young man then, found an odd end of plank lying about and got a carpenter to bore a hole in the middle of it and put in a leg. This was the origin of "Nobody." Ninety Harris used it to sit upon, while he was making the final preparation of clay before rolling it up into lumps to "throw" on to the wheel. He sat at a bench, working the clay up into a paste under the heel of his hand. It had already been trodden, but now the tinier pellets of dryness had to be worked out—for in a pot or pan they would have burst in the burning. So the potter sat picking them out, throwing them in the "squibber" by his side, swaying to and fro, with a pushing motion, "exactly like making up butter," and putting the lumps of clay aside in a heap for carrying to the wheel. Working so, he needed no fixed seat, but this one-legged "Nobody" swayed to his movements, giving him all the support he wanted. No one in the shop had seen such a thing before; but all were glad to use it after Ninety Harris had shown the way. When not in use, "Nobody" lay on its side on the floor.

A conventional stool would certainly not have allowed Ninety Harris or the other shop employees the freely churning motion that made the chore of working up the clay less tiring and hence more efficient—and perhaps even a bit enjoyable. The specialized stools seem to have been affectionately named in recognition of their individuality and preference by the workers, much as some people today are known to name their automobiles. Furthermore, by giving the stools names, the workers could easily demand of a shop boy a particular one.

Sturt goes on to distinguish tools from furniture and apparatus, like stools and the squibber (which was nothing more than a tub of water that gradually filled up with clay), and notes that the potter's tools "were very few." The potters were, however, possessive of what tools they did have. One tool, called a "ribber," was used to groove pots. Ribbers helped make a more uniform groove than could be achieved with the fingers alone, and the implements were frequently fashioned by the potter himself. He so prized them that if he moved to another shop "he did not leave them behind for any successor, but was jealous to take them with him."

Whether apparatus or tools, the form of the potter's equipment developed to make both his gross and fine motions more efficient and reliable, and such are the ends toward which all tools are modified from their unsatisfactory predecessors, thus removing their shortcomings. In his foreword to an illustrated encyclopedia of tools, W. L. Goodman writes that our "progress has been largely a matter of inventing new tools and improving the old ones," but he also points out how frustrating he has found the study of tools, in large part because the tradesmen who "did know and care" about tools did not write about them. Furthermore, a medieval craft was considered a "mystery," and an attitude of secrecy prevailed that has persisted down to modern times:

> A stranger entering the workshop was a signal for men to put their tools away and when any questions were asked about them it was not unusual to offer frivolous or totally misleading answers. As a rule, the men of learning were in no position to disbelieve what they were told and very often the more unexpected the answer the more impressed they were; after all, it was coming straight from the horse's mouth as it were. There are, in fact, several cases where the exact purpose of some tools in common use only a few generations ago are [*sic*] not now known for certain and can only be a matter of more or less informed argument.

There were, of course, some notable exceptions among "the men who wrote the books," including Georgius Agricola (*De Re Metallica,* 1556), Joseph Moxon (*Mechanick Exercises,* 1678–84), and Denis Diderot (*L'Encyclopédie,* 1751–72), but even tools of the last century survive without verbal descriptions of their function, instructions for their use, or names.

The purpose of unusual old tools that collectors have acquired, often precisely because of their oddity and challenge, may be difficult to ascertain, but that is not to say that collectors do not try. One of the delights of antique dealers and collectors, as opposed to the taciturn users of tools, seems to be discovering and explaining the uses of the unusual. One organization of avid tool collectors, the Early American Industries Association, has a Whatsit Committee, and the association's quarterly journal, *The Chronicle,* has a regular column headed "Whatsits?" in which puzzles are posed and solutions proposed for now unrecognizable everythings from scoops to nuts. Look-alike items in old catalogues often present convincing evidence of what something was once used for, but there is by no means unanimity on every mysterious artifact. The purposes of oddly shaped knives and shears are perennial topics of speculation. Some of the most curious tools that have been incontrovertibly identified, and those that we still use today, provide excellent opportunities to test any hypothesis of artifactual evolution, and a universally applicable evolutionary principle that explains how each different tool, no matter how odd, comes to be from its predecessors might assist in identifying the unusual.

Agricola's monograph on mining was one of the first books to record systematically the ways and tools of a craft or trade, and is especially distinguished by its innovative use of illustrations. One shows a silversmith at work on some metal, and into a nearby stump is stuck what looks like a pair of shears, one of the handles of which is bent into an L-shape. It is the bend of the handle that distinguishes this tool from a more ordinary pair of contemporary shears. Agricola thus comments on the oddity and describes it as "an iron tool similar to a pair of shears. One blade of these shears is three feet long, and is firmly fixed in a stump, and the other blade which cuts the metal is six feet long." The term "blade" here includes the handle, and clearly this tool is designed to give plenty of leverage. Its form also serves another function, one that more conventional shears failed to perform effectively.

The problem with using ordinary shears in the situation depicted in Agricola's illustration is that the silversmith has only two hands. If he is without a helper and wishes to cut the piece of metal that he is working on, he has either to put the piece of silver on the edge of the stump and snip it with the shears or to put the silver in the shears and balance it there while the handles are pressed together.

Georgius Agricola's treatise, *De Re Metallica,* was profusely illustrated to show the tools and processes used in mining and metalworking in the mid-sixteenth century. This woodcut shows, among other things, "an iron tool similar to a pair of shears" but differing in having one of its handles bent and formed into a spike so that the shears could be anchored in a stump of wood. Shears not so modified to free one of the metalworker's hands were awkward for unassisted use.

Either maneuver would require contortions, balance, and luck to produce a clean and accurate cut. Thus, ordinary shears fail to be an efficient and effective tool in this circumstance. However, when they were modified, by forming one of the handles into a L-shape, and perhaps by adding a chisel edge at the bottom of the L, the handle could be driven into a wooden stump and the shears could be at the ready to be operated by a single hand while the other steadied and guided the work. No doubt much more efficient and precise work could be carried out with such bench shears, a tool that evolved in response to the failure of conventional shears to work as well as the lone silversmith would have liked.

Specialized tools like bench shears have proliferated throughout history in part because craftsmen necessarily do the same task with the same tool over and over. After a while, the task becomes routine, and the craftsman is able to perform it with predictable skill. The most creative of artisans is frequently one who, in the midst of routine, pays attention to the details of the work and the tools that effect that work, and so it is that the reflective craftsman develops ideas for new and improved tools in the course of working with those that he perceives to limit his achievement or efficiency.

Though modern scholar-craftspeople, such as the television woodwright Roy Underhill or the many talented artisans at Colonial Williamsburg, in Virginia, are more interested in recovering and preserving knowledge of and skill with old tools than in devising new ones, their demonstrations and explanations of the things they work with and on provide much insight into the evolution of artifacts generally. At Williamsburg, many of the tools, especially saws, look just like their counterparts sold in hardware stores today. This suggests that by colonial times these tools had evolved to a high degree of "perfection" for their specialized tasks. Because so many of today's saws have had their form fixed for centuries, we can confidently infer from our own experience with them their use in earlier times.

The earliest metal saws date from the discovery of copper in the Near East some four thousand years ago. As bronze replaced copper, so iron replaced bronze, when each old material failed to be as efficacious as the newer. In the seventeenth century, before wide steel strips could be rolled, the strongest and hardest saw blades were necessarily narrow—hence the widespread use of the bow saw, in which a wooden frame keeps the blade in tension the way a hunting bow keeps a string in tension. Such saws are still popular in Europe, while wide-bladed steel saws have generally replaced the bow in English-speaking countries. Incidentally, this fact, combined with the divergent design of Oriental hand saws, which cut on the pull stroke (as opposed to most Western saws, which cut on the push stroke), provides further evidence that no unique form follows the single function of cutting wood.

The basic principle behind the operation of a saw is, of course, to cut a groove, called a "kerf," into a piece of timber or wood so as to separate it into two parts. The teeth on the earliest saws, said perhaps to have evolved from actual tooth-embedded jawbones of dead animals, must not have been very specialized, but they have evolved into an elaborate diversity of styles, spacings, and settings. Whether wood is cut across or with the grain, for example, presents different problems to the saw teeth, and a single saw with a single edge of uniformly spaced teeth will not work equally well in both situations. In cutting across the grain, individual wood fibers themselves must be cut, and so crosscut saw teeth have naturally evolved into a series of knifelike tools along the edge of the blade. In ripsawing, along the grain, a chiseling action is preferable, and so the ripsaw teeth that have developed to perform that task best look and act like a series of little chisels.

The kerf created by a saw with teeth that are in the plane of a saw blade, as we can imagine they most likely were on the first saws, would tend to fill up with sawdust and pinch the saw blade as it progressed into the cut. Although this failure to function smoothly was corrected in part by spacing the teeth so they double as rake tines to pull out the sawdust while at the same time acting as tools to cut the wood, it was setting the teeth alternately left and right that allowed a kerf wider than the saw blade to be cut and hence not pinch it. But the same saw-tooth design would not have worked equally well for soft- and hardwoods. In cutting the former, a lot of sawdust is produced quickly, and so saws suited for softwood developed wider-spaced teeth with large gaps to catch and carry relatively large amounts of sawdust to the end of the kerf. Hardwoods, on the other hand, yield sawdust much more slowly, and so saws suited to them can and do have smaller and more closely set teeth.

Cutting down large trees naturally required saws that were not limited in the depth of their cut, the way bow saws are. The long- and relatively wide-bladed saw with handles on each end was designed to be operated with the muscle power of two woodsmen, one pulling on the cutting stroke (the long, untensed blade would buckle if it cut on a push stroke) and the other pulling the saw back into position for the next cutting stroke (or cutting also, if the saw had teeth to cut both ways). Once the tree was felled, it could be cut up into logs by the same saw, but then the heavy logs had to be cut lengthwise into boards, and the felling saw failed to do this without encountering new problems. If the log were cut where it lay on the ground in the forest, the saw would have had to be used in a horizontal position, and the saw wielders would have had to stoop very low to perform a difficult task. The four-to-seven-foot length of the saw would have meant that it bent noticeably under gravity, and this distortion, combined with the pinching action of the wood on the blade, would have made it very difficult to get a clean cut. Furthermore, gravity would not assist in removing sawdust. These several negative aspects of sawing boards horizontally out of a log on the ground led to the development of the saw pit, the pit saw, and the pitman.

In order to keep gravity from bending their saw and closing the kerf behind it—and, indeed, to turn gravity to their advantage—the sawyers could respectively position themselves one on top and one below the log, which then necessarily had to have space for a man between it and the ground. This objective was sometimes achieved by propping the log up at an angle or on sawhorses, but in any case this required not only lifting heavy logs a fair distance, but also repositioning them as the sawcut progressed, thus giving gravity the last laugh. For efficient sawing, the entire log had to be raised almost the length of the saw, and the lower man could get his full weight behind his saw stroke only if he could stand up. This arrangement is depicted frequently in Diderot's *Encyclopédie,* but raising a heavy log six or seven feet into the air is no mean task, and substantial sawhorses or scaffolding would be needed to allow the men to resist the undesirable rocking induced in the sawing operation. Whereas this might be the most expeditious way to proceed at a temporary construction site, men who worked a considerable amount of timber at one location came to dig a pit over which the logs could be rather easily rolled into position and manipulated throughout the sawing process. The action of pit sawing is described romantically by Roy Underhill, who seems to relish the opportunity to try any old craft specialty:

> It's a rare music you hear, shuffling ankle deep in fresh sawdust, elbows sweeping scant inches from the tarred plank walls of the sawpit. With each downward stroke the chorus of teeth on the 7-foot-long steel blade rips another half inch along the length of the log. The foot-thick log above your head and the walls of the pit exclude the noise and distraction of the town. There is only the relentless progression of the blade down the charcoal-struck line. . . .
>
> Traditionally, the top sawyer was the senior of the two, owner of the saw and caretaker of its sharpness. . . . The pull stroke of

This illustration from *L'Encyclopédie* shows a two-man frame saw in use to rip boards. The extended top handle relieves the top sawyer from stooping as if to touch his toes with each stroke and keeps his fingers from being caught between the saw and the board, a hazard to which the lower sawyer must be alert. In England and America the bottom sawyer commonly worked in a narrow pit beneath the work piece.

the pitman does the actual cutting of the wood, but he is able to use his weight to his advantage. The top man has most of the responsibility for keeping the cut straight on course and must pull the saw back up with his arms and shoulders alone.

George Sturt, who operated a wheelwright's shop and employed sawyers about a hundred years before Underhill's reveling in the sawpit, had a different recollection of the "sawyers' more recondite work . . . usually under the wheelwright's eye." Sturt's description of the lot of both the sawyer and the pitman was less sanguine and charitable:

Laborious it was in the extreme; and the sawdust poured down on his sweating face and bare arms, and down his back; but at least he was spared the trouble of thinking much. To be sure, he might not go quite off to sleep, although his view went no farther than the end of the saw-pit, and his body and arms were working laboriously up and down for hours. But there were short breaks. Now his mate would call down to him to oil the saw. For this purpose he had a rag tied to a stick, kept in a tin of linseed oil in a crevice of the saw-pit. . . .

But the top-sawyer had no such easy time. He, master of the saw, not only had to keep pace (and more toilsomely, I was assured) with the other's rhythmical lift and pull. It fell also to the top-sawyer to keep constant watch on the work, with a special eye on the saw's action. The least deviation from the straight line might spoil the timber, besides bringing the work to a standstill. And it was more likely to be his saw's fault in bad sharpening. . . .

Sharpening times were bad times for the bottom-sawyer. The temporary rest left him at a loose end for an hour or so. None could blame him if he slouched off for a drink, where he might find a fire to sit by and somebody to talk to. Unfortunately, he was not always in a hurry to go back to work. To the top-sawyer, sharpening was none the more welcome on this account. To know that the other fellow was in the bliss of a tap-room, while he himself was tied to a job, earning no money and using up a sixpenny file—to know all this made sharpening a nuisance at the best. . . .

The sawyers were on the whole so erratic I was always glad to see the back of them. Yet the real trouble was that, as competition grew, a less costly way of getting timber had to be found. At any rate, when planks could be bought in London nearly fit to use, it would no longer do to buy local timber and pay for sawing it, thereby locking up one's money for years while the timber dried. Timber-merchants might do some of that. It was for them to employ the sawyers—or to set up the steam saws.

But long before steam power drove saws down another evolutionary path, sawyers had found fault with the felling saw as a pit saw. Handles set in the plane of the saw blade had worked well for a man felling trees but proved almost impossible to use for one balancing on a log or squeezed into a pit. How much more convenient it was to have the handles set crosswise to the saw blade, so that top and bottom sawyer alike could stand facing the length of the log and watch their work progress without twisting their necks. In time, the top handle came to be attached to a two-foot extension so that the saw could be pulled down to its last teeth without the top sawyer's having to bend over as if touching his toes. And the bottom sawyer's handle evolved into one that was easily removed in times when the saw had to be pulled up out of the pit for sharpening. In every case, the modifications from felling saw to pit saw were made in response to an inconvenience or a failure to perform as well as sawyers could imagine.

Perhaps because the use of saws requires more effort than many other tools, there is an especially great proliferation of styles. Both the felling and the pit saw evolved into excellent tools for their tasks, but they were certainly too large, heavy, and unwieldy to use in the shop of a cabinetmaker or a wheelwright. Thus smaller hand saws developed independently. Some of these were for cutting large planks and panels, but their blades proved still too thick and long to provide the accuracy desired in cutting close-fitting joints and the like. Thus, the back saw evolved with a reinforcing top edge to keep its thin blade from distorting while doing fine work. But none of these saws was suited for cutting curves, and so the coping saw evolved, with its very narrow blade fitted into a frame that does not get in the way. (Because the coping saw's blade is necessarily thin and thus prone to buckling if pushed hard against the work, the blade cuts on the pull stroke like an Oriental saw.) But even this saw fails to work when very tight curves must be cut well inboard from the edge of a board or a panel. Thus, the fret saw developed, with an especially deep bow to its frame and a blade that can cut curves without being angled.

These are just some examples of specialized saws, of course, but they serve to illustrate the evolution within the genre in response to the failure of existing saws to perform a task that arises naturally in the course of timber cutting and woodworking. Failure-driven evolution does not guarantee that each new form will be a success, however, and there are plenty of examples, in old toolboxes and patent files, of failed ideas. One of the problems with the growth of specialized saws was the tendency of users to acquire many separate implements, thus creating problems of capital investment and storage. It must certainly have been the perceived disadvantage of needing separate crosscut- and ripsaws that prompted one inventor to devise a single saw that had crosscut teeth on one side and rip teeth on the other side of its single and otherwise symmetrical steel blade. The handle on this duplex saw was naturally mounted symmetrically on the blade, so that it might serve equally for either set of teeth. Unfortunately, the saw did not work very efficiently, for the highly specialized handle on a traditional saw is set well on the back of the blade and angled for balance and for directing the thrust of the hand most effectively to the cutting edge. By ignoring the details of the highly evolved saw handle, the inventor virtually condemned his duplex model to be a functional failure on all counts and thus an evolutionary wrong turn into a dead end.

One of the most often cited tools in the context of studies of the evolution of form is the ax, and David Pye has cited it as a prime example of form not following function, because to his thinking it may be ideal for producing wood chips but it is terribly inefficient for cutting down trees. Nevertheless, the ax is a tool that cries out to be considered in any theory of the design and evolution of artifacts. Originating in the Stone Age, it developed in effectiveness and form as new materials and new methods of hafting became available. By colonial times, the modern European ax was pretty well established and tradition-bound, and such extratechnological cultural inertia can fix the form of an artifact in its home territory in spite of its inefficiency and functional shortcomings. After all, there is no technological imperative in efficiency, which in any event is in the eye and hand of the tool beholder.

The Europeans had long known that iron is not easily hardened to give a keen edge, and so eventually it became customary to weld a strip of steel onto the business end of the ax head so that the blade could be honed to a fine cutting edge that did not wear out so quickly as iron. But even with a sharp edge, the European ax would tend to twist if not held firmly when being swung, because the bulk of the metal was on the front side of the handle and thus was easily turned aside by any but the most direct and controlled of blows. In young America, however, with its abundance of forests and its pressing need to build houses and other wooden structures on cleared land, the inefficiency and touchiness of the European ax were not accepted so uncritically. As late as 1700, American-made axes still looked very much like their European ancestors, except that they began to incorporate a poll—the blunt end of an ax head that projects out from the back side. The addition of this feature not only made the ax head heavier, and thus capable of packing a bigger wallop, but also moved the center of gravity and percussion back from the blade and closer to the handle, and thus had a stabilizing effect on the ax head during the blow.

By the end of the eighteenth century, the American polled ax was well developed, and it even included a longer cutting edge than its European counterpart. However, no artifact can ever be considered absolutely perfected for all the tasks in which it comes to be employed; an ax design that seemed to be all that one woodsman might want left something to be desired for another. For example, sharpening a dulled ax might not be so inconvenient for the farmer working near his barn or toolshed, where a grindstone was located, but it could be very inconvenient for the woodsman working far away from such a stone. The two-edged ax could thus enable the woodsman to range farther from his home base, because he would have to return only half as often to sharpen his tool.

Many local variations in the ax head developed, perhaps because trees are different in different parts of the country. But there are always subjective aspects to using tools and other artifacts, and, whether by tradition, habit, or feel, each woodsman might be expected to find fault with different and unfamiliar axes and to find one ax to be an improvement over another. In the nineteenth century, there came to be a proliferation of ax styles in America, each varying slightly from the others in its form. The different styles were frequently distinguished by regional names, and may have been selected by users more for chauvinistic than for technological reasons. George Basalla has suggested the range of such localization of form by noting that in 1863 one manufacturer listed felling axes in varieties designated "Kentucky, Ohio, Yankee, Maine, Michigan, Jersey, Georgia, North Carolina, Turpentine, Spanish, Double-bitted, Fire Engine, and Boy's-handled." Within two decades, such a list could exceed a hundred items, each one implying some shortcoming of the others. The name "Kentucky ax," for example, suggests the failure of others to deal with the special problems of the Kentucky woodsman.

Hammers have also been the frequent focus in studies of technological form, perhaps in part because of Marx's astonishment that Birmingham produced so many different kinds. The specialization

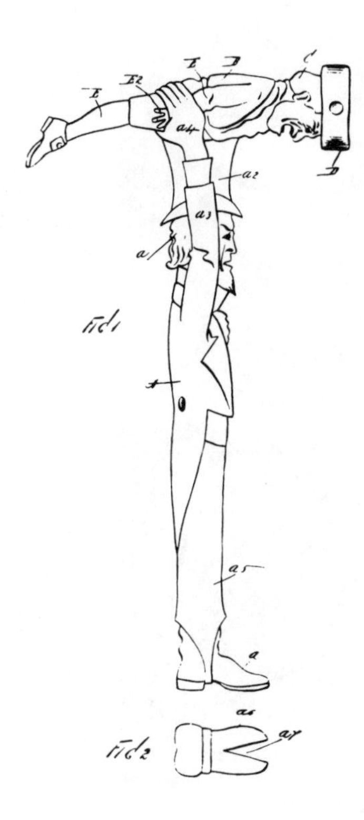

In even the most functional of artifacts, imagination and whimsy can affect form. A design patent was issued in 1898 for this anthropomorphic hammer. Another common tool, a pair of dividers, has frequently been formed with human legs—sometimes very shapely or very muscular ones—by which distances could literally be stepped off at the workbench.

of hammers is perhaps not so surprising when we recognize that, like saws, these tools get a tremendous amount of repetitive use by craftsmen, who thus have considerable thinking time on their hands. Whereas the vast majority of hammer wielders will tend to adapt to and accept the faults of their tools, the rare creative individual will spend time thinking about how to avoid a particular problem that he keeps encountering in the course of using a less-than-perfect tool for a particular purpose. (The observant and inventive toolmaker might also come up with improvements in his products. These even the lumbering craftsman would purchase immediately if he recognized the innovative features as removing a familiar, if unarticulated, shortcoming of an existing tool.)

The astounding number of times a day that a hammer wielder can pound is familiar to anyone who has lived under, or even just in the same neighborhood as, a roof being reshingled. When a roofer once left his hammer behind on my new roof, I was struck by how worn the head was and how polished the wooden handle, at least where it was not taped to contain a crack.

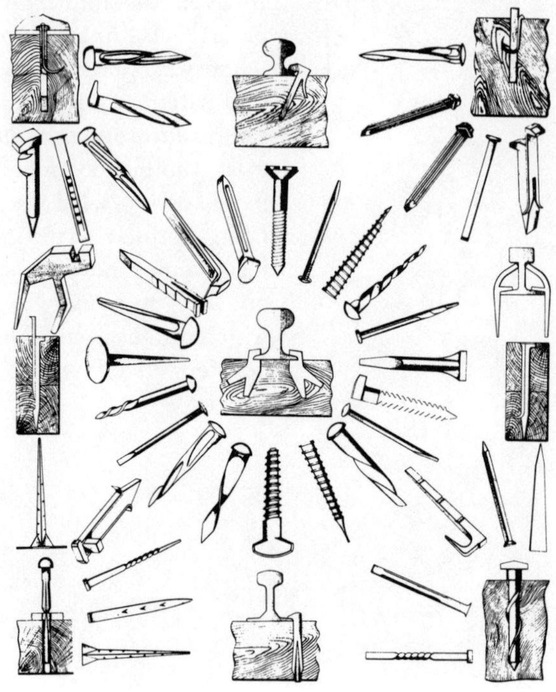

TYPES OF NAILS AND SPIKES.

In the late 1800s, Benjamin Butterworth, nineteenth U.S. Commissioner of Patents, arranged and compiled illustrations from related patents to trace the evolution of artifacts. This page from *The Growth of Industrial Art* shows a great variety of nails, demonstrating the specialized forms of artifacts that can develop out of the failure of existing ones to perform a function as effectively as inventors can envision.

A "collector's handbook" entitled *The Hammer: The King of Tools* contains over a hundred pages of photographs, typically showing ten or twelve different styles per page, of odd and unusual hammers and hammerheads. Another two-hundred-page section of the book reproduces the drawings (four to a page) of U.S. patents granted between 1845 and 1983 for improvements and variations on hammers and hammerlike tools, each one capable of performing some function all other hammers fail to—at least in the mind of the inventor in question. To be sure, many of the variations among tools are covered by design patents, which are granted for novel appearance alone. Still, even though no functional improvement need be present for a design patent, the most successful applications are for artifacts that show "a startling or unique new appearance."

But, for all the personalization and specialization of hammers over the years, late-Victorian merchants were apparently not averse to selling a given style to as wide a range of users as possible. In its 1895 catalogue, Montgomery Ward offered hammers to a strange list of bedfellows indeed, and one light-headed hammer with a large angled claw was clearly to be used in place of other models that defaced baseboards and were incapable of removing stubborn tacks. This was, according to the telegraphic catalogue description,

> The best tack hammer on the market. Will drive tacks without defacing base boards, and will extract the most obstinate tacks without effort. The favorite hammer with upholsterers, carriage trimmers, bill posters, carpet layers, undertakers, photographers, dentists, picture frame makers and cigar dealers. For use about the house it is unsurpassed.

This is a versatile hammer indeed, and one wonders how many hammer styles Montgomery Ward, let alone all of Birmingham, might have had to make, had undertakers and dentists, for example, not been able to work with the same model. That a single hammer could have so many divergent specialized users does suggest that there is a limit to diversity, and that the limit represents a balance rather than a conflict between utilitarian and economic means and ends.

8

Patterns of Proliferation

Among the most talked-about items at antique shows are the odd and unusual pieces of old silverware whose handles clearly indicate they fit a familiar place-setting pattern but whose intended use can be a matter of considerable speculation. Dealers and collectors alike debate function more stubbornly than value, purpose less convincingly than price. The uninitiated does not have to eavesdrop long to experience utter confusion over whether one fine-looking piece is for serving tomatoes or cucumbers and whether another curious item is an ice-cream server, a fish server, or a crumb scoop. The casual onlooker can easily wonder if anyone really knows what he or she is talking about.

Suzanne MacLachlan is a collector of, among other things, any and all pieces of the Vintage pattern of silverplate, which was made from 1904 to 1918 by a division of the International Silver Company under the trademark "1847 Rogers Bros." The pattern incorporated bunches of grapes into the handle design, and so collectors like MacLachlan, who at one time had eleven hundred pieces of Vintage, can with some reason call themselves Grape Nuts. An insurance agent's request for an inventory of her collection forced MacLachlan to catalog her pieces, which led to the publication of her definitive *Collectors' Handbook for Grape Nuts.* The book includes over sixty distinctly different pieces that she had actually seen and acquired, and it contains illustrations of another eighty or so items reproduced mostly from old silverware-dealer and jewelers' catalogues. The pieces range from familiar dinner and salad forks to less common items such as marrow and cheese scoops. The distinctions among knife and fork and spoon can become so blurred that we find in the handbook one hybrid item identified as a "melon knife or fork" and another odd utensil designated an "olive fork or

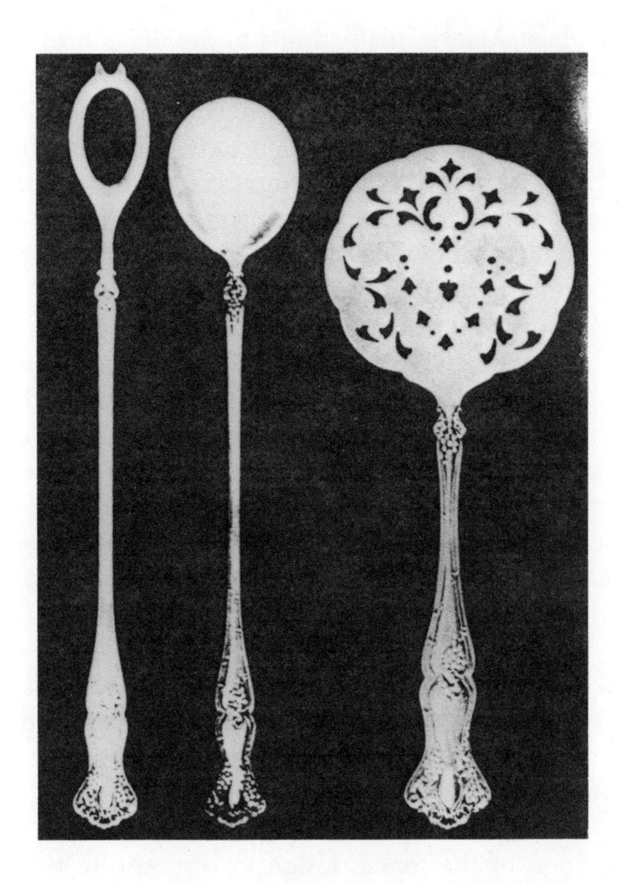

These three pieces of Vintage silverplate are, left to right, an "olive fork or spoon," a chocolate muddler, and a tomato server. Had the olive retriever been made with a conventional bowl deep enough to hold the olive securely, a lot of undesirable liquid might have been transported to the plate. The annular bowl obviates this fault, and the vestigial tines serve to steady an olive before it is scooped up. Neither of the other two pieces would work for olives very well at all.

spoon." This latter is a curious piece whose bowl has two vestigial tines and a large oval cutout in which one can imagine an olive easily resting, and it is a cataloguer's challenge: "An early list has it only as an Olive Spoon, but later lists also include it in the fork column." Other manufacturers were more directly ambiguous and called the patented piece an "Ideal Olive Fork and Spoon."

In her preface, MacLachlan admits to the difficulties of trying to be definitive about the Vintage line: "every time (in two years of compiling notes) that I have written down a firm conclusion from experience, I have been promptly contradicted by additional information!" And later in the book she suggests a source of her difficulty: "A number of factors make grape collecting confusing—but fascinating. From 1904 to 1918 pieces were redesigned, renamed or possibly even changed in size. Special orders combined parts of different pieces, and manufacturers freely swapped blanks, inserts and brand names." The problem is made quite explicit in the case of individual salad forks that carry identical catalogue numbers while having two distinctly different forms of tine. In the earlier model, the tines are wavy and come to rather sharp points; in the later model, the tines are straighter, shorter, stockier, and blunter. According to MacLachlan, this same fork also appears in various catalogues under the names "individual salad fork," "individual pickle fork," "pickle short fork," and "individual relish fork." She further notes:

> Tines on early models are often found badly bent. Tines on later models are straighter and heavier. Rogers' 1847 line carried a lifetime guarantee, and the manufacturers upgraded any piece that required consistent repairs.

The evolution of the fork's design in response to the failure of earlier models to resist bending is a classic example of form following failure. That the evolutionary changes could take place within the fourteen-year lifetime of the Vintage pattern shows how responsive manufacturers can be to the failure of their products to function as intended. However, the change of the piece's name, from salad to pickle to relish fork in the same short period of time, illustrates another, and much subtler, aspect of the evolution of form. Whereas bent tines can be judged by rather objective observation, this other aspect of formal evolution originates in a subjective perception of functional failure, not of the piece itself so much as of related pieces that it is intended to displace. Thus, the salad fork exists at all because the dinner fork somehow failed, or was thought or said to have failed, to function as an effective salad fork, perhaps by virtue of being too heavy or bulky for the lighter chore. The multiplicity of names for MacLachlan's piece suggests its intended multiple uses, for which it was presumably better suited than other pieces.

These forks represent some of the pieces of silverplate once available in the Moselle pattern, including, left to right, a pickle, pastry, and salad fork. The first two pieces show very pronounced cutting tines, which evolved from less robust tines that became bent when forks displaced knives in almost all aspects of eating. The unsymmetrical nature of these forks makes them distinctly right-handed.

The problems faced by Grape Nuts are actually minor compared with what collectors of nineteenth-century patterns can face, since early-Victorian catalogues of silverware manufacturers were not even illustrated. Illustrations were common by the end of the century, however, perhaps because it was nearly impossible to tell some of the pieces apart—with or without a picture—or to identify corresponding pieces in different patterns. Between 1880 and 1900, Rogers Brothers introduced twenty-seven new flatware patterns, which included many new kinds of serving pieces. Other silver manufacturers were equally prolific. According to Dorothy Rainwater, who has written widely on American and other silver:

> By 1898, the Towle Company's "Georgian" pattern included 131 different pieces. . . . There were nineteen types of spoons for conveying food to the mouth, seventeen for serving, ten pieces for serving and carving, six ladles, and twenty-seven pieces for serving that were not classified as ladles, forks, or spoons. One can sympathize with the hostess of that day in trying to be sure that croquettes were not served with a patty server, or cucumbers with a tomato server.

As late as 1926, some patterns were still being made with as many as 146 distinct kinds of utensils. To help simplify the situation for American industry, Herbert Hoover, then secretary of commerce, recommended—and members of the Sterling Silverware Manufacturers Association adopted—a list of fifty-five items as the greatest number of separate pieces that would be in any pattern thereafter introduced. Today it is rare to find more than twenty different pieces in a silver pattern. Confusion about the naming of specialized pieces remains, however, with current catalogues of the best silver companies still calling by disparate names items that look to serve identical functions. Thus, similar-looking pieces of silverware can be called a "cold meat fork" in one catalogue and a "cake or pastry fork" in another, or a "fish fork" and a "salad fork" in still other catalogues. The confusion may be compounded by the fact that differences in form between individual forks within a single pattern are often much less drastic than the differences between, say, what are called "dinner forks" in two distinct patterns.

Such a divergence of form has occurred because the way a standard fork, for example, should be modified to function best as, say, a pickle fork is a matter of some judgment. While a pickle fork is intended to convey the slippery food from a serving container to individual plates, the functional effectiveness of the implement can more easily be criticized in a less elegant context. As anyone who has ever tried to get a pickle from a jar can testify, the standard dinner fork does not do the job very effectively. The tough and

This collection of forks shows the variations available in several silver patterns. *Top row, left to right:* oyster fork-spoon, oyster forks (four styles), berry forks (four styles), terrapin, lettuce and ramekin fork. *Middle row:* large salad, small salad, child's, lobster, oyster, oyster-cocktail, fruit, terrapin, lobster, fish, and oyster-cocktail fork. *Bottom row:* mango, berry, ice-cream, terrapin, lobster, oyster, pastry, salad, fish, pie, dessert, and dinner fork.

slippery skin of a pickle resists even the sharpest tines, although, once speared, the pickle seems quite easily dislodged when passing through the jar's neck. Yet, when a good enough hold is finally gotten on the pickle so that it can be transported to a plate, the fork seems not to want to give up its hold. Should the tines of the perfect pickle fork be shaped to spear and hold the pickle best, or should they be formed to facilitate its release onto the plate? These conflicting objectives, like virtually all the goals of design, demand compromise. Since compromise involves judgment and choice, different designers can be expected to solve the same problem in different ways. In addition, the aesthetic desire to integrate the odd new piece of silverware into a new or an existing pattern can also influence the shape of a fork's tines. Further complications arise when a silver pattern is to have fewer rather than more pieces, for then decisions must be made as to how to fashion single pieces that are to serve multiple functions.

Given the existence of specialized pieces of silverware, the question of what form is for what function may not be an easy one to answer in all cases. Rather than try to do so, many a writer of books on etiquette (as opposed to those on collecting) has suggested that there are indeed more eating and serving utensils than one should care to know about. Emily Post made the point explicitly in the 1920s:

> One of the fears expressed time and again in letters from readers is that of making a mistake in selecting the right table implements, or in knowing how to use one that is unfamiliar in shape. In the first place queerly shaped pieces of flat silver, contrived for purposes known only to their designers, have no place on a well appointed table. So if you use one of these implements for a purpose not intended, it cannot be a breach of etiquette, since etiquette is founded on tradition, and has no rules concerning eccentricities. In the second place, the choice of an implement is entirely unimportant—a trifling detail which people of high social position care nothing about. . . .
>
> The broad statement above, that smart people do not care about which piece of silver to use, has one qualification. They could not use the dinner fork for oysters or a tea spoon for soup, because they instinctively choose an implement suitable for

The form of serving pieces intended for the same purpose could also vary widely from manufacturer to manufacturer, as illustrated here. *Top row, left to right:* sardine forks (three styles), sardine fork and helper, and jelly knives (five styles). *Middle row:* tomato servers (three styles) and tomato fork. *Lower left:* butter knives (four styles). *Lower right, top to bottom:* cheese servers (two styles), cheese knife, and cheese scoops (four styles).

whatever they are about to eat. But whether they happen to choose a medium sized pronged article for fish, that was intended by the manufacturer to be especially helpful for salad or shredded wheat biscuits, makes no difference whatever.

What Emily Post and more recent etiquette writers advise is that a few basic pieces of flat silver are sufficient to set even the best of tables. These essential pieces are: "table spoon, dessert spoon, tea spoon, after dinner coffee spoon, . . . , large fork—often called a dinner fork, small fork—sometimes called salad or dessert fork, . . . , large knife—dinner knife with steel blade, small knife—silver blade, . . ." The ellipses represent specialized spoons, forks, and knives that are included in "the most complete list of flat silver possible in a perfectly equipped home" but that "may be subtracted as unnecessary." But illuminating the origins of the "unnecessary" pieces, which were once no doubt argued by someone to be "necessary," provides valuable insights into the evolution of some familiar if puzzling artifacts.

Much modern silver is very attractive to look at and very comfortable to hold. However, in the course of eating in restaurants and at dinner parties, it is not uncommon to find certain features of a particular place setting that can easily be judged wanting. For example, whereas most dinner forks are generally comfortably large, with four reasonably sharp tines spread over a good width, some forks with obvious pretensions toward modernity have three widely spaced and bluntly shaped tines that are as effective as stumps for eating food. Sometimes, even when a silver pattern's dinner fork is seemingly perfectly proportioned, other forks in the same set can have rather short and stubby tines that make it difficult to spear and to hold securely a piece of lettuce—or anything else, for that matter. Furthermore, with some forks the small surface of the tines, which converge to a shape more like that of a teaspoon than a fork, can offer very little support area and less of a cradle for the food than we might like. As attractive as such forks might be to look at, I have always felt that their business end is not suited to comfortable eating. In short, they seem to be somewhat of a functional failure as forks, yet households and restaurants become committed to them.

Sterling silver is generally a one-time investment, and aesthetics seems very often to play a more important role than function in the initial selection and, often, willy-nilly lifetime commitment to the pattern. Silverplate, on the other hand, is not expected to last indefinitely. Around the turn of the century, better silverplate was sold with the understanding that its plating might be good for twenty-five years with proper care, and it was expected that replating would be done as required. On such occasions, the customer

Emily Post approved of flat silver in this classic pattern. *Left to right:* dinner fork, small fork, oyster fork, dinner knife, small knife, butter knife, fruit fork, and fruit knife. In the 1920s, the popular etiquette writer advocated getting by with few specialized pieces.

could naturally comment, if not complain, about the way a particular piece functioned. Like the maker of Vintage, the responsive manufacturer might realize the advantages of correcting the fault in the next batch made. After all, forks with bent tines could give a whole pattern a bad reputation. But the perfecting of a single piece does not explain the proliferation of specialized pieces.

Whether by blind adherence to tradition or by a tacit recognition of functional overrefinement, Emily Post's Roaring Twenties watchword for choosing silver was conservatism:

> In selecting her silver the bride or householder who would have a perfectly appointed table must be very conservative. Queer pronged spoons, and distorted shapes, whether scooped deep like mussel-shells, or flat-lipped like the petal of a rose, are equally bad. . . .

> The ultimate of perfection . . . is silver that was actually made in the eighteenth or at the beginning of the nineteenth century, because the patine [*sic*] of age is inimitable—to the connoisseur! Happily for most of us, our perceptions are not as keen as the connoisseur's, and we can be very content with modern reproductions that faithfully copy the best originals. . . . Choose reproductions rather than new designs.

But that is not to say that one fork fits all. The large and small dinner forks, for example, evidently coexist because the large fork, so appropriately proportioned for meat, is simply too bold and heavy for more delicate menu items like salad and dessert. The smaller fork, on the other hand, though suited to luncheon dishes, is not substantial and robust enough to be considered ideal for meat. In fact, the small fork that Emily Post illustrates in a set of flat silver she considers "admirable because beautifully simple" is literally nothing but a smaller version of the dinner fork. About an inch shorter but geometrically similar in every way, it meets her criteria for a good fork: "the corners . . . are smoothly round, the prongs are slim." Indeed, according to the arbiter of taste,

> The small fork is the most important fork there is. Its use is for every possible course at breakfast, lunch and dinner except the meat course, for which the large fork is used. The small fork is used literally for *everything* else, and in such great houses as the Worldlys' and Gildings' no other is included in the silver chest.

Even if thought of as a good investment or simply something one is proud to possess, silverware really constitutes a set of tools for the table. Just as specialized woodworking tools proliferated in response to shortcomings that existing tools exhibited when performing a new task, so the pieces of silverware multiplied in response to the failure of existing pieces of silver to perform food-handling and eating tasks at the table as neatly and efficiently as people expected or hoped was possible. Whether customers complained about the trouble they had eating oysters with existing forks, or whether they complained when bringing in for repairs a fork whose tines were bent, or whether taciturn silversmiths saw at their own dining table room for improvement in the way existing implements worked, with time new and modified pieces of silverware clearly did evolve and proliferate. It is certainly possible to imagine that silver manufacturers looked for new pieces to make so as to tempt consumers to buy more, but it is equally possible to argue that something like the Victorian fascination with gadgets and the elaborate meal drove the process.

The silverware that Emily Post takes as a paradigm of design was made during the period when the knife, fork, and spoon were becoming commonly accepted as the basic eating utensils of the privileged classes of Western Europe. Afterward, the size of the knife and fork had alternately grown and shrunk as taste and style in food and utensils argued for larger and then smaller utensils. The successive correction of faults in earlier forks, especially with regard to the number and nature of their tines, and the evolving shape of the knife blade with the displacement of some of its earlier functions by the fork, culminated in the most fundamental forms, if not the sizes, of our basic eating utensils. What followed in the nineteenth century and beyond, however, albeit encouraged by the mechanization of craft labor and the development of marketing instincts and networks, was the gradual realization that what at the time were established as the standard knife, fork, and spoon had real shortcomings at the table. In spite of Emily Post's assertions, it was never easy to eat grapefruit with an ordinary spoon, it was never easy to eat lobster with a large or small fork, and it was never easy to serve asparagus, with any implements. Though it may have been true that the seasoned diner could manage with a few pieces of standard silver, it was also true that those standard pieces failed to function as well as might have been imagined for the increasing variety of dishes that were being brought to the table by technological advances in transportation and refrigeration.

The basic knife, fork, and spoon could no more do everything equally well at the table than could three basic woodworking tools do everything equally neatly in the woodwright's shop. It would seem to be inevitable that specialized eating utensils should have been conceived in response to the frustrations of dealing with the likes of squirting grapefruit, stubborn lobster, and drooping asparagus. With specialization would naturally come the multiplication of eating utensils, to the point where purchasing them could be a financial burden, cleaning and storing them could be a logistical burden, and naming and using them correctly could be an educational burden. Who needed or could afford all of those burdens?

Eventually, with the moral support of the Emily Posts, ordinary people could still feel fashionable without a piece of silver for every culinary occasion. After all, even in the best of houses one needed only some basic pieces.

Yet the nineteenth century was indeed one of gadgets, and none the less so at the dining table. According to one account of fantastic inventions of the Victorian era, when middle-class houses were large and complicated enough not only to entertain grandly but also to care for and store all the things that it required, such entertaining was done by the Joneses "always with the primary object of outdoing their neighbours and acquaintances in hospitality." A formal dinner was frequently the context in which to impress, and one English farmer, "renowned for his generous dining and wining parties," went to elaborate lengths:

> Having invited his friends to dine with him, he objected to the constant interruptions of his servants. He therefore installed in his rambling mansion a railway which connected the dining room with the kitchen and pantry and carried food and wine. An electrically driven car on miniature rails came right to the table through a service hatch and stopped in front of each guest, who helped himself, after which the host pressed a button and off went the train to the next stop, finally disappearing through another hatch on its way back to the pantry to be loaded with the next course.
>
> The mechanical service doll was born of much the same inspiration. A small enamelled figure, dressed as a cook, seventeen inches high and holding in each hand a pan or plate containing food, stood in front of the guest, who pressed a button at the figure's feet and was automatically served.

Although such radical ways of dealing with the servant problem were probably no gravy train for their inventors, the mere existence of these mechanical artifacts, devised to overcome some undesirable aspect of having a meal served, points to the complicated lengths to which the Victorians were prepared to go to improve the way things worked. And such elaborate solutions were not out of keeping with the elaborateness of the meal itself: "The menu at any such self-respecting dinner would include at least two soups, two fish dishes, four entrées, a couple of roasts, two removes and half a dozen as-

A Victorian dinner-table railway answered the objection of a farmer from the south of England to the constant interruption of servants bringing in a meal's many courses. An automaton in the form of a small enameled figure dressed as a cook was patented by another Victorian who wished to avoid as much as possible having real servants in the dining room.

sorted entremets"—i.e., a couple of dishes following the roasts and six different dishes served among the main courses. This would once have seemed quite beyond reason to me, but on a recent visit to England I experienced the vestiges of such extended eating habits. At lunches before I lectured, I was presented with more food in more courses than I am accustomed to eating at all but the most formal of dinners in America. At an ordinary evening dinner in a Cambridge college, I saw more silver than in any American university's faculty club. And at the annual dinner meeting of the Society of Construction Law, held in the hammer-beamed hall of the Inns of Court in which either the first or the second performance of *Twelfth Night* took place (depending upon which conversation I eavesdropped upon), so many different glasses were set that they seemed to form a crystal picket fence down the entire length of the very long table. The varying shapes of the glasses had, of course, evolved and multiplied much as pieces of silver had.

Lest we think that Victorian America had more restraint at the dinner table than Victorian England, a book on social customs published in Boston in 1887 shows that this was not necessarily the case, although perhaps there was some polite concern over overindulgence:

> Seven and even nine wine-glasses are sometimes put beside each plate, but most of us would not approve of such a profusion of wine as this would imply. At other tables, two extra glasses, one for sherry or Madeira, and the other for claret or Burgundy, are put on with the dessert. . . .
>
> After the raw oysters soup is served. At very stylish dinners it is customary to serve two soups,—white and brown, or white and clear. . . .
>
> Fish is the next course, and is followed by the entrées, or "those dishes which are served in the first course after the fish." It is well to serve two entrées at once at a very elaborate dinner, and thus save time. To this succeed the roast, followed by Roman punch [a watery ice containing lemon juice, beaten egg whites, sugar, and rum], and this in turn is followed by game and salad. . . .
>
> Cheese is often made a course by itself; indeed, the general tendency of the modern dinner is to have each dish "all alone by itself" . . . This style, however, may be carried too far. Only one or at most two vegetables are served with one course, and many vegetables make a course by themselves, as asparagus, sweet corn, macaroni, etc.

Such excess was reserved for large dinners, of course, and the same guide to manners assures us that "for a small dinner it is quite sufficient to have two or three wines." In the early years of the new century, eating had become considerably streamlined, at least in the mind of the author of *The Etiquette of New York To-day:*

> Short dinners are the modern fashion. The menu consists, as a general rule, of grapefruit, canapés of caviare, soup, fish, an

l to serve a cutting as well as a spearing function, and fork ting tines were introduced. A "cutting fork" was patented by Reed & Barton. First offered in dinner- and dessert-fork e design soon extended to pie forks and pastry forks and to er cold-meat fork.

e 1882 novel *A Modern Instance* by William Dean Howells, eeper in rural Maine observed a gentleman guest eat mince th a fork as easily as another would with a knife" and admired ill in getting the last flakes of the crust on his fork." Such ity was no doubt aided by the widespread introduction in the of pie and pastry forks with special (for right-handers, at least) g tines, not only widened to resist being easily bent but also ed and flattened to pick and scoop morsels like knives of old. ere also appeared the likes of salad forks, lemon forks, pickle , asparagus forks, sardine forks, and more, each with its tines ned, thickened, sharpened, splayed, barbed, spread, joined, or ehow modified to reduce the faults that other forks exhibited in lling some very specific food. But not all forms of forks evolved irectly, and though the knife may have been endangered in the ing years of the nineteenth century, it was no extinct species. cial dishes would continue to frustrate the diner using existing, eit multiplying, utensils.

The vast differences in texture between fish and roast make them spond quite differently to the knife and fork. Properly cooked fish kes readily, of course, whereas meat need not. But many foods ve divergent responses to the knife and fork, and so this alone ems not to have been a sufficient reason for the standard dinner nife and fork to fall out of use for fish, and so for specialized implements to evolve. Yet it became common in the late nineteenth entury for etiquette books to assert that fish especially was never o be eaten with a knife, although, in the style of the genre, the books generally offered no explanation for the prohibition. By the early twentieth century, the specialized fish knife and fork had come to be standard tableware, but still few explanations were offered as to why the table knife of the day could not be used.

To this day, writers on etiquette seem at a loss to explain exactly how the oddly shaped fish knife is to be used, and Emily Post considered it "wasteful, since it must be bought and kept polished for no other purpose than for eating fish." Even if this had become true in the 1920s, there must have been perceived shortcomings of the

entrée, a roast with two vegetables, game and salad, dessert and fruit.

Cheese is sometimes served after the game. If artichokes or asparagus are served they are separate courses.

Though it may be little wonder that to service such meals a plethora of specialized pieces of silverware evolved in the nineteenth century, wonder does not explain form. Even a plot to sell as many pieces of silver as possible would not in itself explain why the individual pieces looked the way they did. What does explain their form is the failure of the elements of the standard place setting to perform as efficiently as imaginable the great variety of cutting, slicing, piercing, scooping, and other operations that would be required to eat a great variety of foods. Since there were so many courses, it was necessary to set the table initially with a sufficient number of implements, or to bring clean ones with each new course. There were naturally many times during a meal at which used silverware was properly taken away with the used china, and there was clearly a need to keep things moving as smoothly as on a railroad, lest an evening's dinner last into the next day:

> In order to give an elaborate dinner it is almost indispensable that one should have a large quantity of china and plate, otherwise the delay from washing the dishes will be endless. . . .
>
> When one plate is taken away at the end of a course another is at once substituted for it. If a knife and fork are laid on this, the guest should take them off promptly, otherwise he may delay the serving of the next course.

Washing silver between courses would have been, as it still is, inconvenient at best, and so a great number of individual pieces of silver would naturally have had to be owned by anyone who wished to entertain in a grand way. Such a multiplicity of silverware could have been achieved by having a great many identical knives, forks, and spoons of standard design, of course, but this would not have obviated the failure of the common knife and fork to work as well with, say, fish and shellfish as they did with a piece of roast.

The oyster fork, for example, appears to have evolved from the standard or even the small fork because the latter's tines were so long and gently curved that they could not easily work oysters whole from their more deeply curved shell. The older fork could have been used like a lever to pry the oyster free, of course, but this might risk launching the morsel across the table. The oyster fork's short tines allow the leftmost one to be used as a blade to sever the oyster from the shell, the fork's small curved tines allow it to conform to the shape of the oyster shell, and the fork's shorter handle allows the diner to exercise better control over this delicate action. The end tine of oyster forks also came to be used for scooping out from the shell the meat of lobsters and the like. This kind of action, along with that of severing stubborn oysters, might have caused the cutting tine to bend over time, and so it came to be widened while at the same time retaining its thinness through the depth of the bowl (so that it possessed a cutting ability) and a pointedness (so that it had some efficacy as a lobster pick). Regardless of their thickness or pointedness, closely spaced tines could get in the way when one was attempting to reach into a lobster claw to extract meat, and so many a seafood (née oyster) fork came to have its tines more widely spaced, or even splayed out, to facilitate the action. As fashion and tastes changed, designers groped for the optimal form, not only for aesthetic reasons but also to eliminate functional failures.

Amid the proliferation of specialized tableware in the latter part of the nineteenth century, the general form of forks was still more or less set. Yet arbiters of taste were very sensitive to the use of even the standard fork, apparently because, having only relatively recently evolved its fourth tine, it was still among the newest table tools of an increasing segment of the population. A book on social customs published in 1887, for example, followed a brief history of the fork's emergence with a caveat:

> All English-speaking nations, however, as well as the French, now absolutely forbid the use of the knife except to cut with. On the Continent, society is not so strictly divided by the "knife line;" and it would not be safe in Germany, for instance, to judge of a man's social position by his method of using his knife. . . .
>
> The fork has now become the favorite and fashionable utensil for conveying food to the mouth. First it crowded out the knife, and now in its pride it has invaded the domain of the once powerful spoon. The spoon is now pretty well subdued also, and

the fork, insolent and triumphan
tyrant. The true devotee of fashion
except to stir his tea or to eat his sou
ice-cream with a fork and pretends

A contemporaneous writer on "mode
of her readers might be interested in kn
the use of the knife is of comparatively
universal among civilized people:

> In England and her colonies, and in Fra
> ica the "knife line" is strictly drawn; bu
> those who adopt the French manner
> Swedes, Italians and Germans, often th
> their mouths and do not consider it inele

Another writer cautioned that all "made dis
rissoles, patties, &c., should be eaten with a *fo*
should not be used in eating them, as a knife w
and out of place; it would therefore be a vulg
But, with so much favoring of the fork, it ha
at the table, and even a few specialized forks co
all things equally well. And, to make matters w
admitted that the fork was "much more difficult
knife." Given a social climate that encouraged th
new gadgets, many specialized forks came to be
shortcomings of existing ones multiplied with t
menus and the concurrent diminished use of the
The pastry fork, for example, came into existence i
new fashion, described in 1864 by the etiquette wri
as "foolish" but "fashionable":

> It is an affectation of ultra-fashion to eat pie with a fo
> a very awkward and inconvenient look. Cut it up firs
> knife and fork both; then proceed to eat it with car
> in your right hand.

What Emily Post termed "zigzagging" thus appears to
practiced in the mid-nineteenth century, but in time
dropped completely out of the act of eating pie. Hence

alone ha
with cut
in 1869
sizes, th
the larg
In th
an innk
pie "w
his "sk
dexter
1880s
cuttin
point
Th
forks
wide
some
hand
so d
clos
Spo
alb

standard knife and fork for eating fish that caused the fish knife and fork to evolve as they did. Understanding the technological context in which this occurred enables us to understand better why the "wasteful" utensils have the form and use they do.

It was the acidic nature of fish, often aggravated by the addition of lemon juice, that prompted a change in table manners and that ultimately led to a new form of tableware. Acidic fish juices corroded the steel from which knife blades were generally still made in the late nineteenth century, silver being too soft to take and hold a keen cutting edge. The book *Manners and Rules of Good Society,* which was written by an anonymous "Member of the Aristocracy" and which was in its thirty-third edition in 1911, indicated that a table knife and fork had indeed long been used to eat fish, but that it was the failure of the steel-bladed knife to work satisfactorily in the aggressively acidic environment that prompted change:

> It was then discovered that a steel knife gave an unpalatable flavour to the fish, and a crust of bread was substituted for the knife. This fashion lasted a considerable time, in spite of the fingers being thus brought unpleasantly near to the plate, and to this day old-fashioned people have a predilection for that crust of bread. One evening a well-known diner-out discarded his crust of bread, and ate his fish with two silver forks; this notion found such general favour that society dropped the humble crust and took up a second fork. This fashion had its little day, but at length the two forks were found heavy for the purpose and not altogether satisfactory, and were superseded by the dainty and convenient little silver fish-knife and fork which are now in general use.

The fish knife and fork were being provided "at all formal dinners" by the late 1880s, according to another writer, who also noted that the old rule against using a knife with fish "was so very inconvenient, especially in eating shad." The new knives were of silver, of course, because that did not corrode in the fish's acid, and they were most notably "of a peculiar shape and of small size, as also are the forks that accompany them." The peculiar shape of the fish knife, which might be described as notched-back scimitar, appears to have evolved in part out of the failure of the dinner fork to deal effectively with a whole fish on the plate. The head and tail would have had to be ripped rather than cut away, and a general ripping of the skin would have had to be performed to get the flesh off the backbone. All the ripping would have naturally left a lot of loose bones to be dealt with in the fish and thus in the mouth. Even though a knife of silver would not be so sharp as a steel knife, its blade would certainly be keen enough to sever the head and tail and slice a properly cooked fish along its backbone. The blade did not have to be long to reach into the fish and separate the meat from the backbone, but a wider-than-normal blade was very effective in keeping the fish from flaking apart and sticking to the bones. The odd notch near the knife tip was evidently a vestige of the fork tines that had necessarily performed these operations before, and might also serve to catch the backbone, once this was loosened from the fish, and keep it from slipping off the knife while being swung to a place on the plate away from the fish meat. To distinguish the silver fish knife from the more common steel-bladed ones, the more ornamental blade also evolved. The fish fork, which worked in concert with the knife in performing these operations, also functioned better for having proportions wider than those of a normal fork, since this made it less likely to fragment the fish in the deboning operation.

Emily Post may have considered the "special fish-fork wasteful" and declared that "fretwork trimmings across the prongs are absolutely taboo"; still, she did admit that "the plain [fish] fork with a flattened first tine, and the silver knife with the pointed end and saw-tooth edge, are not taboo, because their designs have tradition." But these rather recently specialized utensils really owed their form and existence more to technological adaptation than to a tradition spanning mere decades. And though the introduction of stainless-steel knife blades in 1914 may have made even the silver fish knife itself a bit "wasteful," by then its highly specialized if mysterious form had firmly displaced the ordinary knife for fish; silver had become the "traditional" material for the fish knife. As much as she may have eschewed functional explanations for including one piece of silver and excluding another in her chest of necessities, in her own dining Emily Post may herself have recognized the downright aptness of the fish knife and fork, even if she could not bring herself to articulate it except in terms of "tradition."

Other specialized pieces of tableware, whether considered traditional or not, have also evolved because they removed the inconveniences and worse that arose in employing the usual pieces in unusual situations. Thus, the fruit knife and fork, the former with a pointed tip and sharp edge, and the latter often with three very sharp tines, reduce the spray of fruit juices at the table and make piercing and slicing fruit much more convenient. The grapefruit spoon, pointed to match the shape of a fruit segment and serrated at the tip or along one edge to assist cutting out the pulp, has a great advantage over the teaspoon that is immediately evident to anyone who has squirted or been squirted by someone across the breakfast table. The iced-tea spoon, also variously called a lemonade or ice-cream soda spoon, has obvious advantages over the teaspoon in the tall glasses in which the cool drinks are served. In the early part of this century, these spoons were made with hollow handles that doubled as straws. Whether or not spoon and straw should be in the glass simultaneously, they often were, and anyone who has tried to avoid being poked in the eye by the spoon, or to keep the wet straw away from the spoon end, will immediately recognize the convenience of the invention.

Emily Post may have had the innate wisdom to eschew the folly of the Victorians when she declared unnecessary the large number of specialized pieces of silver that had evolved from the classic place setting, but her reasoning was a bit askew. The newer pieces themselves were not without function; indeed, they enabled the *fin-de-siècle* diner to eat an elaborate meal with a style and good form that some late-twentieth-century observers of society and culture would love to regain.

The multiplicity of pieces of tableware exhibited in such early-twentieth-century patterns as those collected by Grape Nuts fell out of fashion with faster times, harder times, and smaller homes. In 1965, for example, Reed & Barton's "Francis I" pattern was advertised with only "the ten most essential serving pieces" out of the original seventy-seven pieces it had contained in 1907. Silverware patterns now typically contain a fraction of the special pieces of a century ago, with many pieces doing multiple duty, and there still appears to be little standardization in the forms or names of the utensils. What looks like a fish fork in one pattern may be called a salad fork in another, and vice versa. What is offered as one pattern's butter knife looks curiously like another's fish knife, though it is perhaps a bit smaller. Confusion appears to abound, and in some of the more modern patterns especially the pieces can't be identified without a catalogue—if then.

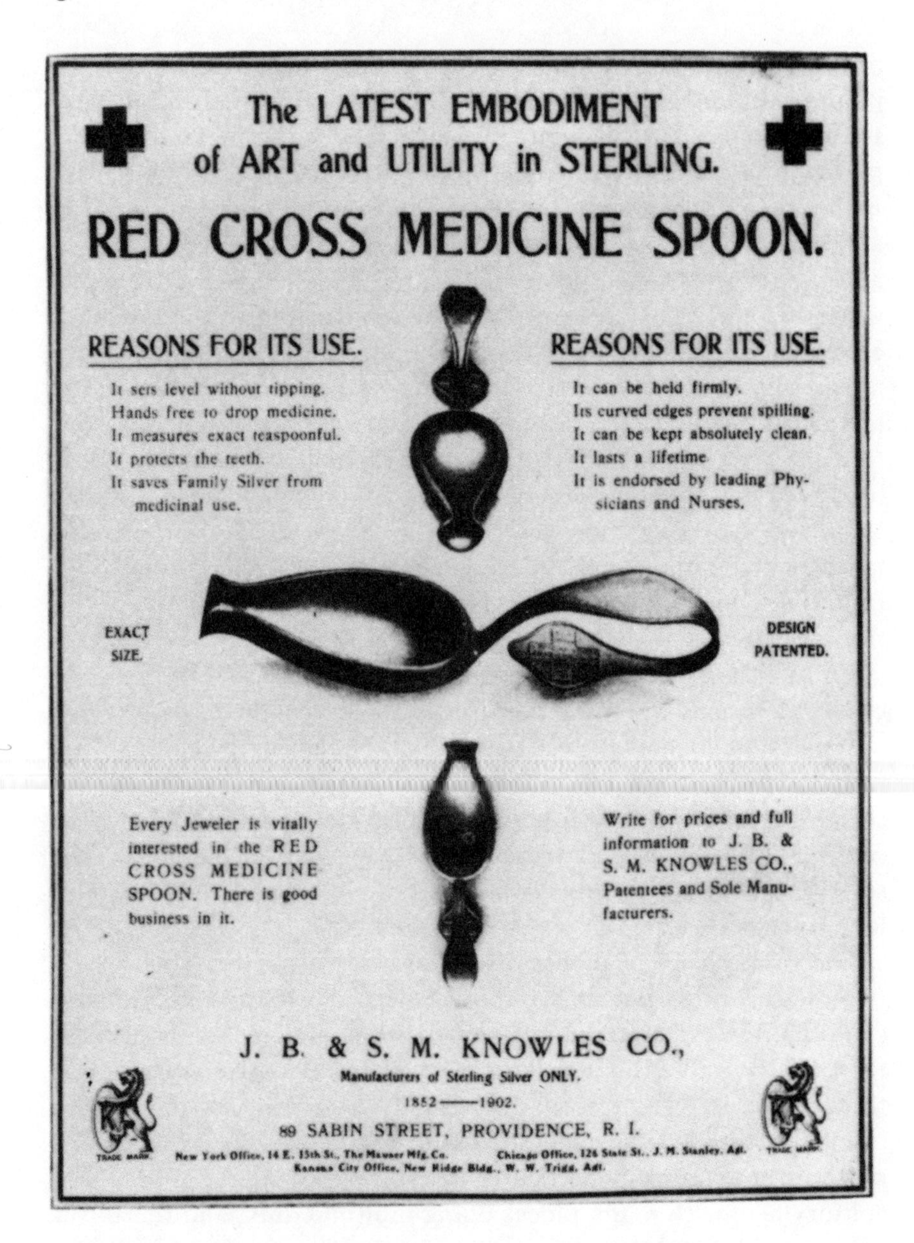

Many specialized pieces of silver were developed in the Victorian and Edwardian periods, including this design-patented medicine spoon that answered to many of the real or imagined objections parents had to using a good teaspoon for a sick child. Patents also protected processes for reinforcing silverplate at points of heavy wear on forks and spoons.

Many of the most contemporary silverware patterns appear to be designed more for how the pieces look than for how they work, and this would appear to contradict every rational expectation of technological evolution. But the paradox is resolved if we understand that there is a kind of design that can ignore function entirely. We might say that this is a "form eschews function" school of design, and one that places considerations of aesthetics, novelty, and style above everything else. Whatever existing silverware there might be, it can always be argued by purveyors of the new to be less balanced, less new, and less stylish than the latest offering. The knives and forks that come out of such design considerations can often appear to have their blades and tines growing organically out of the handles, where the unity of the pieces originates, and from which inspiration seems to spring. But to design from the handle is to shoot from the hip when it comes to silverware, for the business end of the individual pieces is where the action is going to be. Though Emily Post may not have perceived that tradition emerges out of the minimization of failure, there is no excuse for a designer to overlook the fact. Yet this is exactly what modern product designers seem to do when they strive so hard for a striking new look that they throw out function with tradition.

9

Domestic Fashion and Industrial Design

A chef's knife and a joiner's saw perform similar functions in analogous contexts. Each is used by a frequently sullen artisan to prepare the parts of some grand design, whether it be an elegant dish for the table or a fine sideboard for the dining room. Since cooking and joinery are ancient arts, the business ends of cutting tools have evolved to a highly specialized state, and different knives and saws are used according to the task at hand. But whether the handles on a chef's set of knives or a joiner's collection of saws match or are attractive is seldom the overriding feature by which they are chosen or upon which the artisan's talents or work is judged. Rather, a master's favorite old knife or saw may have so chipped and splintered a handle that no apprentice would likely ever choose it over a newer model. The visibly misshapen handles of many long-used tools neither recommend nor fit them to any but the craftsman whose hand has eroded them over a lifetime as imperceptibly as a river does its canyon's walls.

A table knife also shares functional traits with kitchen knives and wood saws, but the social context in which the table implement is used places it in a different category entirely. There is an element of social intercourse present at the dinner table, where actions are steeped in the conscious and unconscious traditions and superstitions associated with breaking bread, that is simply not present at the kitchen counter or the workshop bench. There the artisan works by and large silently and alone, amid a creative disarray of parts and tools. In contrast, the diners around a table are seldom creating anything but conversation and the other ephemera of a dinner party—a performance in the round in which they are both actors

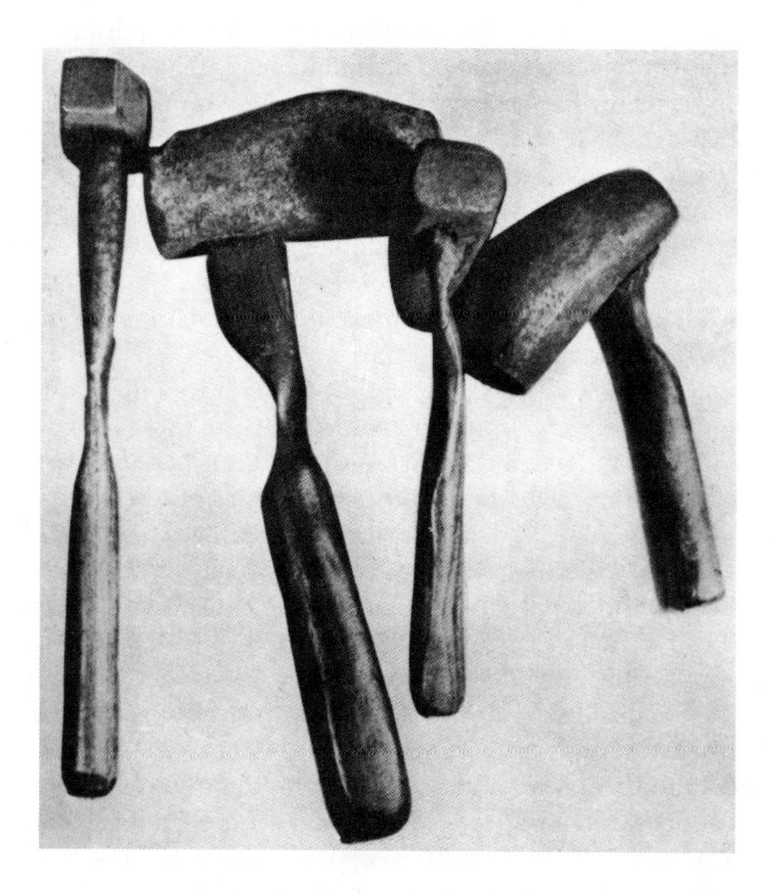

These hammers, ranging in size from about seven to eleven inches long, exhibit wooden handles whose once-straight lines were eroded over the years of constant use given them by nineteenth-century Sheffield craftsmen pounding out small work on cutlery. The uneroded parts of the handles show them to have been of a common shape when new, and the different patterns of wear can be attributed at least as much to each workman's individual grip as to the grain of the wood.

and audience. Indeed, the most essential thing that does take place at a dinner table is not supposed to be creative, but is, rather, expected to conform to the often arbitrary rules of manners, etiquette, and fashion.

The consumption of food, like the wearing of clothes, is something we all do. When these things were done by our primitive ancestors, they may have paid less attention to style than to substance. But with the advance of civilization, including in particular the development of class distinctions and the emergence of mass production, the ability to make and the desire to own a variety of things in a variety of prescribed styles came together in the mixed blessing of a consumer society. The social context in which an artifact is used can indeed have a considerable influence on the more decorative and nonessential variations in its form. However, the evolution of functional details is still very much driven by failure in contexts ranging from the genial to the sullen.

In spite of Marx's astonishment that five hundred different kinds of hammers were made in Birmingham in the 1860s, this was no capitalist plot. Indeed, if there were a plot, it was to not make more. The proliferation of hammer types occurred because there were then, as now, many specialized uses of hammers, and each user wished to possess a tool that was suited as ideally as possible to the tasks he performed perhaps thousands of times each day, but seldom if ever in a formal social context. I have often reflected on the value of special hammers while using the two ordinary ones from my tool chest: a familiar carpenter's hammer with a claw, and a smaller version that fits in places the larger one does not. The tasks I've applied them to have included driving and removing nails, of course, but also opening and closing paint cans, pounding on chisels, tacking down carpets, straightening dented bicycle fenders, breaking bricks, driving wooden stakes, and on and on.

When I use my ordinary hammer for something other than driving or pulling nails, I normally do not do a very good job; the damage that I inflict on the object of my pounding suggests a modification of my hammer for that special purpose. In closing paint-can lids, for example, I have learned to pound carefully if I do not want to dent the top and make it difficult to get an airtight seal; a hammer with a very broad and flat head would be better. In pounding on chisels, I have noticed that my hammer often slips off or misses its mark; a very large-headed mallet would be better. In tacking carpets close to a baseboard, I have either gouged the baseboard, bent the tack, or smashed my thumb; a long and narrow head, magnetized to hold a tack in place, would be better. In trying to straighten out dents in a contoured bicycle fender, I have found that even my smaller hammer has too large and flat a head; a ball-peen hammer would be better. In attempting to break bricks in two by striking them with my hammer's claw, I have gotten slanted edges at best; a hammer with a chisel claw set more nearly perpendicular to the handle would be better. In pounding wooden stakes into the ground, I have found it difficult to keep a stake's end from splitting; a hammer with a broader and softer head would be better. In short, if I were doing these things not only now and then on weekends but every day on a job, I would want just the right hammer to do the job just right. If I were to try to accomplish five hundred different things with a single hammer, I might find at least five hundred faults and invent more than five hundred variations of the hammer. And as with the hammer, so with the saw and other tools; the quality of my work and my reputation could suffer if I did not have the proper specialized tools.

Whatever my profession, my social reputation rests more on how I handle silverware than on how I do a hammer. But highly specialized pieces of cutlery have now fallen out of fashion, and so eating with the few that remain can be even trickier than hammering. Since the days when diners brought their own knives and forks to the table are long gone, we are expected to adapt instantly to whatever odd and unusual piece of silverware might be set before us, whether or not its end fits the food and whether or not its handle fits our hand. This state of affairs is as much a result of the evolution of manners, style, and fashion as it is of the rational development of form. Indeed, the latter can actually be curtailed by the external factors of economics and the arbitrary clock of fashion.

When Emily Post advised her 1920s readers to eschew any but the most traditional silverware, she held up the classic patterns of the late-eighteenth and early-nineteenth centuries as models of good taste. When she went further and stated that silver actually made during that period was the epitome, she effectively forced herself to speak ill of the many specialized knives, forks, and spoons that had evolved since the classic silver was made. It was not simply a question of what could be afforded, for surely anyone to whom antique silver was affordable (something "possible only to the longest purse") could also supplement it with modern implements done tastefully in the same style. No, it appears to have been the ultimate in consumerism—having what the have-nots could not have—that Emily Post captured in her "Blue Book of Social Usage," by declaring that antique silver was the only real silver. Although by the twenties almost any socially conscious host or hostess could buy and "be very content with modern reproductions that faithfully copy best originals," only old money or a great deal of new money could hope to possess the real thing.

People no doubt read into Emily Post's declarations and assertions what they could afford to read, and those with short purses might have looked not to etiquette books but to mail-order catalogues when chosing their silverware patterns. In a 1907 catalogue of the "very best English goods," silver forks in the Old English, Queen Anne, French Fiddle, King's, and other patterns are indistinguishable until one looks above the roots of their tines. Indeed, if a sheet of paper were used to cover all but the tines of these forks, the catalogue reader could not tell if they were of the same or different patterns. An array of six silver fish-carving sets shows the same phenomenon; except for their handles, each five-tined fish-serving fork is indistinguishable from the others. Two columns of knives show some variation in the blades, with the pointed tip being reintroduced apparently more for style than for purpose, and blades are more distinguished not by their shape but by their decorative etching. Pickle forks and butter knives are among the few specialized pieces pictured. Prices are given for a variety of sizes of forks and spoons and a limited number of serving pieces, but none are pictured, presumably because they differ only in size and not in form from the standard table fork and spoon. However, the proliferation of patterns was no insidious plot, for the consumer was expected to choose only one of the many, and the manufacturer and the merchant actually had to tie up considerable capital in a widely varied stock. Indeed, the need to offer a multitude of choices was imposed on the merchant to keep the customer from going elsewhere for a pattern chosen for the faddish detail at the end of its handles rather than the functional details on its functional tip.

Notwithstanding the consolidation that occurred within silverware patterns after 1926, catalogues for the American masses, contemporaneous with Emily Post's pronouncements about selecting reproductions over new designs, show a variety of patterns that she no doubt had in mind when she observed that "on bad silver the fork-corners are sharp, the prongs thick, and something is added to, or cut away from what is supposed to be a plain design." Sharp corners and thick tines made forks less easily bent at the table and in the sink of the servantless household, but the thick tines also made the forks less effective to spear food. Such paradoxical developments arose from the focus on choosing a silverware pattern whose supposedly functional ends followed in line with their fashionable handles rather than adhering to the classic designs that had evolved in response to the failure of their predecessors effectively to separate food on the plate and convey it to the mouth.

Whatever the shape of its tines, making a fork out of a more common metal, such as britannia (a compound of tin, copper, and antimony that surpassed pewter in luster and hardness), and plating it with a fixed amount of silver produced an affordable piece of flatware. If thicker tines and more elaborate handles added surface area to be plated, then the plating could be thinner. The often heavily decorated handles that distinguish one silverplate pattern from many others were clearly the focus of illustrations in mail-order catalogues. Frequently the identical bowls of spoons were overlapped as the handles fanned out in a display of choices in the same price category. Elsewhere, handles only were pictured in an arrangement of patterns radiating out from a text announcing "quality" and "charm" and assuring the shopper that "no effort has been spared in making this service correct, individual and charming." Individuality seems to have been an important selling point, with engraving offered either free or at a nominal charge, and the guarantees of at most a lifetime suggested that the silver was not intended to be bought for one's descendants but to express one's own individuality.

In the wake of the government-sponsored simplification system, there appears in such catalogues also to be a diminishing emphasis on specialized pieces of silver, such as oyster forks and fish knives, and a growing emphasis on serving pieces, such as sugar shells and gravy ladles. This was in keeping with a return to dining *en famille* after the late-nineteenth-century vogue of *dîner à la Russe,* in which the entire dinner was served in the Russian fashion from a side table, with no serving dishes appearing before the diners. Today, the dexterous use by waiters of the large fork and spoon in concert, rather than the employment of specialized serving pieces, reminds us of how versatile a few pieces of silver can be in practiced hands.

Even in catalogues of some of the best modern silver, handles are much less likely than blades or bowls or tines to be obscured or omitted from an illustration. In an exhaustive catalogue of silverplate patterns intended for collectors, nothing but handles are shown, as if to stress that even to the practiced eye there is little to distinguish knives, spoons, and forks of one individual pattern from another. Designers of tableware were certainly not in complete agreement that knife blades, spoon bowls, and fork tines had evolved to perfection; any designer who thought about it could no doubt have come up with some slightly different solution to the problems of existing cutlery. But it appears that, early in this century, as opposed to the gadget-prone Victorian era, eating implements had clearly become objects more of fashion than of function.

Where fashion does not monopolize form, it is the business end of a tool that gets the most attention. Thus, in a collectors' handbook of hammers, handles are consistently cropped from the photos of at least a thousand unique tools. And in a book on country craft tools, one illustration of a wide variety of hammers shows several with their handles cut off, and among those handles that are drawn complete there is very little variation compared with that of the heads. Indeed, the illustration raises the question why the handles have not become as specialized as the heads; the answer may be that craftsmen are more interested in how their tools affect the work than in how suited they are to the hand.

The greatest variation in hammer handles appears to be in their length, a feature related more to the magnitude of the hammer's blow than to its grip. Another illustration of hammers, taken from the exhibition on materials at the National Museum of American History, shows a wider variety of heads but a similarly limited range of handles. There are some unusual handles, notably those made of metal, but there is certainly no attempt to individualize them. This may be in recognition of the fact that no two hands are the same. Besides, the craftsman's hand will soon adapt to the handle it works with as surely as we adapt to the handles of the silverware set before us. There is little room for style on the workbench itself.

The relationship between fashion and form—or, rather, the influence on the latter of the former—did not escape Staffordshire potters in the eighteenth century. Josiah Wedgwood was one of those potters, and he wrote in his experiment book about how traditional stoneware was priced so low that "potters could not afford to bestow much expense upon it, or make it so good in any respect as the ware they would otherwise admit of; and with regard to elegance of form, that was an object very little attended to." Of imitation tortoiseshell in particular he wrote that, since "no improvement had been made in this branch for several years, the consumer had grown nearly tired of it; and though the price had been lowered from time to time in order to increase the sale, the expedient did not answer, and something new was wanted to give a little spirit to the business." But even though the desire to sell more pottery was a clear objective, this is not to say that modifications in its form were made arbitrarily for fashion's sake. Wedgwood sought to generate business not by the mere fact of novelty or specialization but, rather, by the elimination of shortcomings in conjunction with fashion. Since "people were surfeited" with existing products, Wedgwood wanted in any change "to try for some more solid improvement, as well in the body as the glazes, the colours, and the forms of the articles of our manufacture."

Wedgwood's constant experimentation with evolving form and style was motivated by scientific curiosity directed to the elimination of defects as well as to marketing strategy. The scientist Wedgwood was elected to the Royal Society in recognition of his substantive research into matters of the kiln. But throughout much of his long business association with the Liverpool merchant Thomas Bentley in designing, manufacturing, and marketing ornamental wares like vases and urns, Wedgwood was reticent about advertising the significant technological innovations that ultimately made the now famous neoclassical designs possible. Precursors to the successful designs, which were accepted in no small part because neoclassicism was the fashion of the times, were not so widely embraced by the consumer, and it required a certain sense of correction of failure, whether advertised or not, before capitalism was rewarded.

Writing about style in architecture, the nineteenth-century theorist Viollet-le-Duc asserted that "style consists in distinction of form," and complained that animals expressed this better than the human species. He felt that his contemporaries had "become strangers to those elemental and simple ideas of truth which lead architects to give style to their designs," and he found it "necessary to define the constituent elements of style, and, in doing so, to carefully avoid those equivocations, those high-sounding but senseless phrases, which have been repeated with all that profound respect which most people profess for that which they do not understand." Furthermore, he argued that the theory becomes clear only in the example: "If ideas are to be communicated, they must be rendered palpable and tangible. If we wish that style, as regards form, should be comprehended, we must consider form in its simplest expressions."

Viollet-le-Duc takes the example of coppersmithing, "one of the primitive arts," and considers an early copper vase made with only an anvil and hammer by a workman paying attention to obviating failure:

> His first care is to make a flat, circular bottom for his vase, in order that it may stand firm when it is full; and to prevent its contents from spilling when it is moved, he contracts its upper orifice and then spreads it suddenly at the edge, to facilitate pouring out. [This] presents the most natural shape, given by the method of fabrication, for such a vase. To enable the vessel to be lifted conveniently, the workman attaches handles to it by means of rivets; but, as the vase requires to be inverted when it is empty, that it may be dried, he so shapes these handles as not to arise above its upper level.

A vase fashioned in this way has style, according to Viollet-le-Duc, but what he suggests—namely, that the vase is fabricated in the coppersmith's first rational attempt to do so—is highly unlikely. Moreover, some of the details of form that the theorist argues from function would seem to be debatable in their interpretation. For example, it might make more sense to have the handles project slightly above the top, perhaps thickening them to protect against their bending, so as to allow some air to get under the upturned vase and thus assist in drying it. Indeed, the vase that Viollet-le-Duc describes is really an intermediate stage in the evolution of the form he has chosen to study. But, in spite of beginning *in medias res,* he does go on to show how the form can change first for the better and then for the worse:

> But the coppersmiths themselves, in their desire to do better or otherwise than their predecessors, soon quit the line of truth and propriety. There comes then a second coppersmith, who proposes to modify the form of the primitive vase in order to seduce the purchaser with the attraction of novelty; to this end he gives a few extra blows of the hammer and rounds off the body of the vase, which until then had been regarded as perfect. The form is in fact novel, and it becomes fashionable, and every-

Viollet-le-Duc used the design of copper vases to illustrate his ideas about style. *Left to right:* "the most natural shape," with handles that are not likely to be bent when the vase is inverted for drying; a modified form, with a more rounded bottom, "to seduce the purchaser with the attraction of novelty"; and a still rounder form, from a "capricious and fanciful" designer seeking greater novelty, with handles that are susceptible to being bent in use.

body in town must have one of the vases made by the second coppersmith. A third, seeing the success of this expedient, goes still further, and makes a third vase, with rounder outlines, for anybody who will buy it. Having quite lost sight of the principle, he becomes capricious and fanciful; he attaches developed handles to this vase, and these he declares to be in the newest taste. It cannot be overturned to be drained without danger of bending these handles, yet every one applauds the new vase, and the third coppersmith is regarded as having singularly perfected his art, while in fact he has only robbed the original work of all its style, and produced an object which is really ugly and comparatively inconvenient.

Individual aspects of Viollet-le-Duc's argument may be debated precisely because different critics and designers will see different shortcomings in the vase and will perceive different solutions in its form. This is why there are seldom only three designers involved in such an evolutionary train, especially when one comes up with something novel and fashionable that "everybody in town must have." Some might prefer the third vase's shape, and a fourth coppersmith, for example, might easily correct the fault of the handles' being bendable by making them heavier in a way consistent with the other lines of the vase. Or he might make a poorer design, thinking he is strengthening one feature but in fact weakening another, which it will take a fifth coppersmith to improve upon. Or a sixth designer, perhaps finding the reinforced handles too heavy aesthetically, would lighten them again. Though such modifications might have appeared to Viollet-le-Duc or others to have been grossly inferior, each in turn might have become the rage of consumers everywhere and in its time might have been the definitive vase to copy. *De gustibus non est disputandum,* but in the twentieth century a new breed of designer would call upon taste to account for itself.

Industrial design as an explicit and public marketing tool, rather than an unnamed and unspoken component of business-as-usual in the off-limits corners of many a factory, did not really come into being, at least in America, until the Great Depression. The self-proclaimed originator of the field was Raymond Loewy, who arrived in New York in 1919 as a young man in the uniform of a French army captain. During the 1920s, he worked mostly as a free-lance illustrator for fashion magazines and upscale department stores like Saks Fifth Avenue and Bonwit Teller. Through a friend, who introduced him to lunches at the Algonquin Hotel and summers on the New England coast, he made the acquaintance of many sophisticated New Yorkers.

In 1927, while working on ads for the original Saks, on 34th Street, Loewy was invited by the company's president, Horace Saks, to visit the uptown site where a branch store was being readied. Loewy expressed his opinions about how a department store should be, in today's terms, an integrated system. The employees should be selected for their "physical appearance and courteous inclination," and they should be well though simply dressed. Elevator operators, with whom the shoppers would willy-nilly become "quite intimate" during crowded rush hours, should be "correct, polite, and neat" and should wear uniforms. The store's wrapping paper, boxes, bags, and other details should be attractively and consciously designed, and there should be a unified advertising campaign to introduce the new store. This system was a tremendous success, of course, and so was Loewy. But he was not satisfied with the course of his career as a fashion illustrator, even though the Depression left few other opportunities for his talents. Loewy was an observer not only of society but also of its products, and even before the Depression he did not like all that he saw, which was plenty.

So many functionally similar consumer products had evolved that the principal competitive shortcoming they seemed to face was a failure to distinguish themselves one from another. Since they could not easily do so in their operation, they tried to do so in their fashion. Thus, different brands of toasters were distinguished by superficial features and fashionability. Yet this was not necessarily exploiting the consumer, for no one was expected to buy more than one toaster. Rather, each manufacturer vied for any competitive advantage to attract the consumer who needed or wanted a new toaster. But something was wrong, according to Loewy, who admitted that, "with few exceptions, the products were good." He was "disappointed and amazed at their poor physical appearance, their clumsiness, and . . . their design vulgarity." He found "quality and ugliness combined," and wondered about "such an unholy alliance":

> Once in a while, a product would be more cohesive in its design. But then it would be utterly spoiled by a lot of applied "art": a mess of stripes, moldings, and decalcomania curlicues that would hopelessly cheapen the product. It used to be called gingerbread. (Now we call it schmaltz, or spinach.) What's more, all this corn was expensive: it did not generate spontaneously; it had to be painted on, etched in, stamped out, slid over, pushed out, or raised up; baked in, sprayed, rolled in, or stenciled up. It meant unnecessary work and, therefore, parasitic cost increase to the consumer. I was shocked.

Loewy was also "shocked by the fact that most preeminent engineers, executive geniuses, and financial titans seemed to live in an aesthetic vacuum," and he believed that he could "add something to the field." But, not surprisingly, the people he approached were "rough, antagonistic, often resentful," and, by his own admission, Loewy's French accent was not so helpful outside the fashion world.

However, he believed that creation of consumer demand was part of the solution to the Depression, which was compounded by a form of fear that manifested itself in "a lack of imaginative products and advanced manufacturing," compared with what had been. Loewy was but one of the more conspicuous and self-promoting of "a few industrial design pioneers who were able to make some business leaders aware that this lack of vision and industrial timidity" was not something that was good for business, and "success finally came when we were able to convince some creative men that good appearance was a salable commodity, that it often cut costs, enhanced a product's prestige, raised corporate profits, benefited the customer, and increased employment."

Among the first to be convinced was Sigmund Gestetner, a British manufacturer of office duplicating machines, who met Loewy on a visit to America. In 1929 the Gestetner machine looked like an ungainly piece of factory equipment, with an exposed pulley and drive belt and four protruding tubular legs that provided support and stability but had little else to recommend them. According to one of Loewy's accounts, he was asked if he could improve the appearance of the machine and he answered, "Certainly." After agreeing to a fee, he had a hundred pounds of modeling clay delivered to his apartment and went to work. According to another of Loewy's accounts, Gestetner was not so easily sold on a redesign of his machine, and Loewy got the job only after sketching a stenographer tripping over one of the protruding legs, sending papers flying everywhere. Regardless of the genesis of the commission, Loewy essentially redesigned the machine by eliminating some of its failures: by cleaning up its awkward lines, by making its cabinet of warm wood rather than of cold metal, by covering up its ugly pulley and belt, and preventing accidents by making the legs flush with the body of the duplicator. The model change was introduced later in 1929 and, according to Loewy, "is generally considered the first American example of industrial design before industrial design was understood as a conscious activity."

What seems to have overcome any initial reservations that Gestetner may have had about allowing a virtual stranger to redesign the appearance of the duplicating machine was Loewy's drawing of an objectively indisputable failure—the secretary tripping over a protruding leg. Gestetner was convinced there was a problem to be solved, but there is no indication that the solution affected the quality of copies made by the machine. Other manufacturers seem to have been persuaded in similar ways that they needed an industrial-design consultant. Loewy described the typical prospective client of the 1930s: "He makes nice Widgets, they sell all right, and he doesn't believe he really needs any help from the outside." Loewy won him over by pointing out to the manufacturer problems that he had hardly realized he had:

> Your present models seem to lack certain physical characteristics that would make them stand out among the competition. For one thing, they might reproduce better in your newspaper advertising. The present models are rather weak in appearance and they lack sparkle and highlights. We feel that a competent outside organization with design imagination, working in close co-operation with your engineers, might develop a fresh and unusual answer to your problem.

It was easier coming up with fresh and unusual answers to some problems than others, of course, and Loewy admitted that this affected the fees his firm charged. Redesigning a big thing, like a tractor, commanded a relatively low fee, because "there are so many obvious things one can do to make it better-looking," but he would charge a very high fee to redesign something like a sewing needle. The key was identifying the problems with an existing design and proposing changes. Certainly even a well-evolved design like a needle has problems, such as its propensity to prick the finger and its ability to resist being threaded. But the finger can be protected with a thimble, and the eye threaded with a wire device, and thus the needle's sharp point and small eye have been preserved so that the instrument can perform its primary function of sewing effectively. What Loewy might have created that was new and unusual he did not say, perhaps because no needle manufacturer was willing to pay him a $100,000 fee to solve a problem with which sewers had long learned to live.

Tailors and seamstresses had also come to expect pins and needles to be packaged a certain way, and they expressed little if any need for a change. But industrial designers like Loewy seem to love to redesign much familiar packaging, often pointing out the problems with the old only in the context of the new. In his memoirs, Loewy illustrates his 1940 redesign of the Lucky Strike cigarette pack, for example, with before and after photos. The old pack was basically dark green, with the familiar brand name in a target on the front and a description of the toasted tobacco blend on the back. According to Loewy, the green ink was expensive and had a slight smell. His redesign removed these problems by making the package basically white and moving the "It's toasted" slogan to the side. The word "cigarettes" was put in much smaller and discreet type, supposedly because the brand name and shape of the pack alone conveyed what it contained. The red "Lucky Strike" target was placed on both the front and the back of the pack, so that discarded packaging was always lying right side up and advertising its brand to passersby.

Loewy's ambitions were not only to design small packages, however; from boyhood he had loved railroads and their locomotives. Having obtained a letter of introduction to the president of the Pennsylvania Railroad, Loewy was greatly disappointed that his lack of experience designing railroad equipment got him little more than a polite "we'll call you" at their meeting. In desperation, he pleaded with the president, "Can't you find one single design problem to give me now, today?" When asked what he had in mind, Loewy responded, "a locomotive." The young designer's hubris apparently prompted an impish response from the president, and he gave Loewy the opportunity to redesign the trash cans in Pennsylvania Station.

Loewy was ecstatic to have any railroad commission, and after studying the use and abuse of the existing trash cans he came up with sketches of new designs. Several prototypes were built and tried out in the station, and soon he was called back to the president's office. Loewy asked repeatedly, "How's the trash can?" but got no response. The president seemed to want to talk about everything but trash cans. When finally pressed, he told Loewy, "in this railroad, we never discuss problems that are solved." He then called in the man in charge of locomotives, who showed off photos of an experimental one that was soon to be built in quantity, and asked Loewy, "See anything wrong with it?" He did, of course, and thought to himself, "It had a disconnected look; component parts did not seem to blend together, and its steel shell was a patchwork of riveted sections. It looked unfinished and clumsy." But with the designer in the room, Loewy said only, "It looks powerful and rugged," and that he thought it could be "further improved." What he did was sketch his ideas and recommend that riveting be replaced by welding, at a savings of millions of dollars in fabricating costs, and so the first streamlined locomotive came to be made. However, the growing propensity of Loewy and other industrial designers to streamline everything from toasters to pencil sharpeners soon suggested that a failure to be fashionable more than a failure to function was often dictating form.

Within two decades of the first repackaging of the Gestetner duplicating machine, industrial design was firmly established. Writing of the postwar years, Loewy claimed that "no manufacturer, from General Motors to the Little Lulu Novelty Company, would think of putting a product on the market without benefit of a designer." Whether an employee of the firm or an independent consultant, the industrial designer seemed to "know what the public wants." And although Loewy was perhaps the most flamboyant of the new breed, his focusing on problems with existing designs was not unique.

Henry Dreyfuss was involved with designing theater sets in New York before opening an industrial-design office on Fifth Avenue in 1929. His influence on the appearance of things from John Deere tractors to Bell System telephones gained him a considerable reputation, and many an aspiring designer sought his advice. To one inquiry he responded with an exercise to help assess talent and aptitude, and it centered on identifying problems with existing designs:

> Walk through a department store or carefully examine a mail-order catalog or just look around your own home. Select a dozen items that do not suit your fancy and seriously study them, then make an attempt to redesign them.

Dreyfuss assumed the individual had some art, architecture, or engineering training, and that there was a degree of self-confidence, along with an ability to accept objective criticism of any redesigns offered to the master. Though appearance was the most obvious and often most easily criticized feature of an existing design, Dreyfuss was a strong advocate of what have come to be called human-factors considerations, and in his book *Designing for People* he laid down a five-point formula for good industrial design. Admitting that other designers might not state the case exactly as he did, Dreyfuss was still convinced that his five points comprised the essential concerns of the whole profession. The points are: (1) utility and safety, (2) maintenance, (3) cost, (4) sales appeal, (5) appearance. In ascending order, these points appear to become further removed from basic function, but they all can serve as criteria for how various aspects of failure in existing things can be improved by redesign.

One thing that has resulted from the emergence of industrial design is the proliferation of artifacts that are competing for attention by declaring themselves "new, improved," or "faster," or "more economical," or "safer," or "easier to clean," or "the latest," or whatever comparative (or superlative) suggests or asserts that one product is better than its predecessor or its competition. But there is also an apparent reluctance among consumers to accept designs that are too radically different from what they claim to supersede, for when familiar things are redesigned too dramatically the function they perform can be less than obvious and thus possibly suspect. Loewy summarized the phenomenon by using the acronym MAYA, standing for "most advanced yet acceptable." Dreyfuss emphasized the importance of a "survival form," which was manifested in "a familiar pattern in an otherwise wholly new and possibly radical form," thus making "the unusual acceptable to many people who would otherwise reject it."

Industrial designers thus seem to know not to go too far too soon in making changes, no matter how rational these might be. According to John Heskett, in his study of industrial design, practitioners have learned to "strive for a delicate balance between innovation in order to create interest, and reassuringly identifiable elements." What determines the expected form of anything is a kind of fashion. And fashion more than function is without question what determines so many of the contemporary forms that surround us, whether they be on the highway, the workbench, or the dinner table. But a myopic obsession with fashion, whether in silverware or steel bridges, can lead to the premature extinction of even the most fashionable form if it does not anticipate failure in the broadest sense, including the failure to be fashionable tomorrow.

10

The Power of Precedent

An interesting example of a multitude of forms solving the same functional problem occurred in pottery making in the late seventeenth century. Whether by whimsy or wisdom, there came to be made a curious category of earthenware known as "puzzle jugs." These devices had odd projecting tubes, hollow handles, and hidden conduits that carried the liquid in deceptive and unexpected ways when the jug was tipped to the mouth. If the drinker did not figure out how to drink from the jug, it acted much like a practical joker's dribble glass. Such achievements of the potter's art were not beneath even the famous Wedgwood family, and, according to a nineteenth-century biographer of Josiah Wedgwood, there was plenty of reason to design variations on the basic puzzle or problem, which was to make it difficult to drink the contents of the jug without spilling any:

> It became a prolific source of wagers, and most ale-houses found it to their advantage to keep one or more of different forms for their visitors. The handle usually sprang from near the bottom of the jug, and was carried up its "belly" some distance, when it bowed out in the general form, and was attached to the rim at its top. The handle and rim were made hollow, opening into the inside of the jug near the bottom, and around the rim were attached a number of little spouts, differently placed, according to the whim of the potter. The ale could thus only be drunk by carefully covering up with the fingers all the spouts but one, and through this one the liquor would have to be sucked into the mouth. Beneath the handle a small hole was, however, usually made, through which, if not carefully and closely covered, the ale would spill, and thus cause the discomfiture of the drinker and the loss of his wager.

The jugs themselves were often inscribed with mottoes and verses to taunt the drinker. For example, one jug read,

From mother earth I took my birth,
Then formd a Jug by Man,
And now stand here, filld with good cheer,
Taste of me if you can.

Another offered:

Here, gentlemen, come try yr skill,
I'll hold a wager, if you will,
That you Dont Drink this liqr all
Without you spill or lett some Fall.

And still another put it this way:

Gentlemen, now try your Skill,
I'll hold you Sixpence, if you will,
That you dont drink unless you Spil.

The variety of taunting verses demonstrates a range of literary solutions to the same verbal problem: to communicate a lighthearted challenge to the jug user. This nonuniqueness in the way language can transmit a single idea is also suggestive of how various forms can accomplish the same function. Indeed, the variety of verses inscribed on puzzle jugs was surpassed by the variety of jugs themselves. In addition to those with tubes protruding from every which place, there were jugs pierced through the middle, jugs with an inside tube passing from the handle down to the bottom, and jugs whose double sides included infundibular cores. The variety illustrates well that no unique form followed the single function of outwitting the drinker. Though it might be argued that the function of these vessels of practical joking was deliberately to have a deceptive form, the very fact that this could be accomplished in so many different ways serves to underscore the options available to designers, and the fun designers can have. Whereas a broad variety of solutions is generally not sought for production in a typical problem of product design, in the creation of puzzle jugs there was a definite premium on a puzzling array of forms. And their designers clearly

Earthenware "puzzle jugs," such as these two examples, were produced by the Wedgwood family in the late seventeenth century. These ale jugs were deliberately designed to be confusing to use and served as a basis for wagering in alehouses. The drinker would bet he could down the ale without spilling any, but to do so he had to cover up the right combination of holes and tubes, lest the jug behave more like a dribble glass. Had a unique form existed, the practice of wagering might not have been so popular.

had little trouble coming up with a bewildering variety of solutions to the same problem: how to trick the drinker into dribbling.

Not all artifacts are designed to trick the user, of course, and user expectations regarding form can actually constrain designers. By the end of the nineteenth century, the configuration of a standard bicycle, like the motorcycle of today, had achieved a rather mature form, which has not changed substantially since. The turn-of-the-century bicycle worked rather well in its context, and the kinds of modifications that have taken place generally have involved mechanical improvements in brakes, gearing, and tires rather than any dramatic alterations in the way the frame, wheels, handlebars, and seat fit together. This is not to say that the bicycle had evolved into a technologically predestined form, for cycling enthusiasts and designers have long found the old standard balloon-tired workhorse wanting in speed and efficiency (and have come up with designs that put the rider in positions ranging from recumbent to prone). Rather, the two-wheeler that we all might sketch if asked what the archetypal bicycle looks like is what has come to be the accepted and expected form of compromise among the competing things a bicycle is expected to provide: inexpensive, speedy, reliable, and relatively comfortable transportation that is faster than walking but less tiring than running.

But nothing is perfect, of course, and one of the failings of a bicycle might be said to be its requirement that the rider be also its power. This is fine for trips of moderate distance over manageable terrain, or for people looking for exercise along with or even over transportation, but there are clearly situations where a source of power other than human legs is highly desirable. Thus a problem with the bicycle could easily become a problem to design a motorized bicycle or, more concisely, a motorcycle. Although the problem of designing a motorcycle may be prescribed in the positive terms of fitting a motor to a bicycle to give the new vehicle advantages over the old, in fact the problem derives rather directly from a criticism of the existing device, from the failure of the bicycle to go under its own power. The formulation of the design problem is but a structured articulation of the objective of removing a shortcoming from an existing design.

The very articulation of a problem, such as "fit a motor to a bicycle (so that the rider can be transported faster and more effortlessly)," can strongly suggest a solution. In practice, the nonverbal conception of a solution by an inventive mind is often what prompts the inventor in retrospect to articulate the problem and couch it in the language of a need. After such a rationalization of a nonrational leap of creativity, what remains is how to effect a solution in a way that minimizes objections and introduces fewer inconveniences than it removes.

The cover illustration for the issue of *Science* containing Eugene Ferguson's insightful article on nonverbal thought in design showed but eight possible turn-of-the-century solutions to the problem of motorizing a bicycle. Not only had the motor somehow to be connected to a wheel by a drive mechanism, but a fuel tank and possibly a battery had to be fitted to the bicycle frame. These ideas might have come in a flash of inventiveness, but, as the illustration so graphically demonstrates, what the motorcycle would look like depends very much on how the component parts were fit together. Assuming they are all technically feasible, the eight configurations taken two by two can best be compared by identifying their individual advantages and disadvantages, which are like opposite sides of the same functional coin. Form may be said to follow function only in the sense that heads or tails follows each flip of the coin. The gaming analogy goes only so far, however, because, unlike the gambler, who is bound by the final coin toss, the designer in the end may pick and choose retroactively which tosses to bet upon in the marketplace.

Among the many imaginable combinations and permutations of the components of a motorcycle, one places the motor away from the rider, thus eliminating any potential interference with the legs. But locating the motor behind the bicycle requires an extension of the frame, thus increasing the cost of vehicle and altering its center of gravity. What constitutes the "best" solution among the various candidate designs is a matter of judgment and compromise; in the final analysis, the detailed form of the motorcycle does not follow its function in any predetermined way, but ultimately rests on a judgment of which choice is least undesirable. What might ultimately come down to an arbitrary choice among competing configurations, as manifested in the location of the fuel tank, for example, in time can become so strongly associated with motorcycleness that, even if functionally relocated in a new (and improved) design, a vestigial tank (a "survival form") may be retained in what has become the customary location. The design critic John Heskett has noted a striking example:

> The Ariel "Leader" motor-cycle, . . . produced in Britain in 1957, had a petrol tank located on the rear frame, but retained a dummy-tank of conventional form. This same device was later repeated on the Japanese Honda "Gold Wing 1000", the dummy tank opening in half to reveal electrical controls. In both cases, producers felt unable to present a visual choice to the consumers in face of the power of the conventional image of a motor-cycle, even though its form had become functionally redundant.

Just how far one detail of a design problem can affect form is also illustrated by a more recent "radical innovation" in motorcycle design, in which the powerful motor (now an "engine") is so large as to itself serve as the frame to which wheels, seat, and other equipment are directly attached. This recalls the form of early motorized tractors, in which the engine and transmission casting also served as the frame to which axles, steering wheel, and other barest essentials were attached. A simple iron saddle seat was mounted directly over the transmission, and the driver's feet rested on small stirruplike protuberances, thus giving the impression that the horseless machine itself had been harnessed and was being straddled and ridden much like a living steed. Before that, one of the first steam-powered tractors was actually hitched to a team of horses, not for their power but because there was as yet no mechanical means of steering the machine.

One of Raymond Loewy's early commissions was to improve the design of an International Harvester tractor, which even as late as 1940 appeared to be little more than an engine on wheels covered by the barest of protection and having a steering linkage that looked remarkably like a rein. The tractor's high seat was difficult to reach without getting a leg up, its iron-cleated wheels were prone to clogging with whatever mud they did not throw up on the exposed driver, and the tricyclelike arrangement of wheels made the entire machine rather unstable in tight turns. Loewy's improved design gave the machine four rubber-tired and spokeless wheels, fenders, and the beginnings of a streamlined body that evoked the form of an automobile more than that of a horse. What Loewy did for International Harvester's tractor, Henry Dreyfuss did for John Deere's, and though the two share similarities with what has come to symbolize "tractorness," each also had its distinctive silhouette.

Everything designed has an element of arbitrariness in its form. Loewy described how groups of his designers used to go about designing a new model automobile. Different groups were given different tasks, such as the front and rear of the car, and the conceptual work began, to be cut off at some predetermined time by deadlines that were imposed at the outset. After a time, there were "piles of rough sketches," and Loewy saw the design proceed as follows:

> Now the *important* process of elimination begins. From the roughs, I select the designs that indicate germinal direction.

> Those that show the greatest promise are studied in detail, and these in turn are used in combination or arrangements with one another. A promising front treatment can be tried in combination with a likely side elevation sketch, etc. From this a new set of designs emerges. These are then sketched in detail. After careful analysis, they boil down to four or five.

The form of the final design continues to evolve through full-scale plaster or wood mock-ups, and even at this stage there can remain a degree of arbitrariness; "when several models are to be shown, it is advisable to paint them all the same color so that color preference will not influence the choice of the management unduly." Choices are made not to scheme against the consumer but to choose what would seem to be the best design and therefore the best bet for recouping investment in research and development:

> Changes are inevitably suggested and another complete showing is arranged to demonstrate how these alterations have been incorporated in the design. When the final okay for production is given the design cycle is complete. It is up to the engineering and production departments to draft it and detail it.

The detailing of a design involves translating the final management decision into precise drawings and specifications so that the thing can be produced. Though designers and engineers can present a multitude of solutions to the design problem, and can argue technologically, aesthetically, and economically for this one over that, it is seldom an engineering decision alone that determines the look that comes off a production line. In cases where the roles of engineer and manager are combined in one person, he or she must wear different thinking caps at different times.

Loewy gave a further illustration of the absence of predestination in design by relating the story of his involvement in a lawsuit over patent rights in which a client of his was suing another manufacturer for design infringement. According to Loewy, it was "a clear-cut case" in which the competitor simply copied the appearance of a product Loewy had designed. The defense argued that the design patent was invalid, because "the product could not possibly be designed any other way and still function properly." The case had been dragging on for weeks when Loewy was called as a witness for the client. In the exchange that ensued, the attorney asked Loewy if the particular product could be "designed in any other manner and still be practical and function properly," and whether he could do so. When he answered positively, Loewy was asked if he could demonstrate such alternative designs, and he replied that he could, by making some sketches. He was then asked to do so, and, according to his own report:

> I unfolded my easel, placed the drawing board on it, and started making rapid sketches in large black outline, visible to anyone in the back row. Ten minutes later, I had about twenty-five designs, all different, most of them attractive, all of them practical.

Loewy's ego and business interests seem to have led him to stress his successes, arbitrary as their form might have been; in the final analysis, the design selected would have been the one that, in some compromise way, failed least to satisfy both designer and client. The plurality of solutions to a given problem—and their shortcomings—are virtually inescapable in design.

Designers less gregarious than Loewy, and working with less conspicuous things than locomotives, have tended to call themselves not designers but inventors. Lyndon Burch, an inventor of circuit breakers, electromechanical switches, and waterproof thermostats that enabled electrical appliances like frying pans and coffee makers to be immersed for washing, got his first real break when he was hired as a design engineer by a New Jersey thermostat manufacturer, which obviously expected him to solve problems associated with the company's business. According to his own description of how he thought about problems, his mind worked basically with shape and pattern:

> Most of my work really involves geometry—simple geometric structures to perform a function. So I'll start with a geometric pattern in my mind. . . . After I see the pattern, I'll try to find fault with it, and nine times out of ten, I *can* tear it to pieces, so I'll start again. But when I've got the right pattern, somehow I just know it's right.

Burch clearly was able to come up with tentative solution after tentative solution to the same problem. Even if he could tear 90 percent of his solutions apart, this is not to say that they did not solve the problem. They simply did not solve it as well as he imagined they could, or they did not show to Burch the promise he was looking for. For example, one of his important inventions of the late 1940s was a metal switch for thermostats. Existing switches had worked on the principle of a disc of metal responding to temperature changes by snapping through from one position to another, following much the same principle that causes a metal noisemaker to respond to thumb pressure, or the recently faddish slap bracelets to curl around the wrist in a snap. Burch dismissed variations on familiar devices and came up with the idea of achieving a large movement in response to a small one by cutting a flat piece of metal into various shapes that responded in a twisting fashion to pushes and pulls. Thus, the same function, that of responding in a big way to small influences, could be accomplished in a new manner, and this enabled manufacturers to make (and patent) new switches and thermostats which they could claim accomplished functions similar to the snapping disk without infringing on the patents of others.

All patents contain explicit "claims," which are often seemingly interminable sentence fragments following the colon ending a rubric such as, "What is claimed is," "We claim," or "I claim." The claims come at the end of a patent, and they ostensibly lay out exactly what is being patented. According to patent attorney David Pressman, claims say to the public:

> The following is a precise description of the elements of this invention; if you make, use, or sell anything which has all of these elements, or all of these elements plus additional elements, or which closely fits this description, you can be legally held liable for the consequences of patent infringement.

Pressman, giving do-it-yourself advice to the independent inventor who wants to write his own patent application, not only instructs the reader in the basics of writing the sentence fragments of claims but also gives, under the heading of "other tricks in claim writing," the advice to "use 'weasel' words like 'substantially,' 'about,' or 'approximately' whenever possible" in specifying a dimension, for example, "to avoid limiting your claim to the specific dimension specified." Pressman also explains why "many patent attorneys recommend that a claim not appear too short":

> A claim that is short will be viewed adversely (as possibly overly-broad) by many examiners, regardless of how much substance it contains. Thus, many patent attorneys like to pad short claims by adding whereby clauses, providing long preambles, adding long functional descriptions to their means clauses, etc. The trick here, of course, is to pad the claim while avoiding a charge of undue prolixity.

The legal implications of patents may encourage technical writing at its worst, but the phenomenon is nothing new. In a 1906 patent of theirs, Orville and Wilbur Wright, through their attorney, list eighteen claims for their flying machine The first describes what we today would call one of the wings of a biplane, but then it was called what would become the name for the entire machine:

> In a flying machine, a normally flat aeroplane having lateral marginal portions capable of movement to different positions above or b[e]low the normal plane of the body of the aeroplane, each movement being about an axis transverse to the line of flight, whereby said lateral marginal portions may be moved to different angles relatively to the normal plane of the body of the aeroplane, so as to present to the atmosphere different angles of incidence, and means for so moving said lateral marginal portions, substantially as described.

One of the few things that this claim makes unambiguous is that the aeroplane, or wing, was "normally flat"—i.e., plane—in the Wrights' early conception of a flying machine. They and others would eventually discover, of course, that a cambered wing would provide more lift and thus make the double aeroplanes of biplanes unnecessary and, incidentally, the word "aeroplane" (now "air-plane" in the United States) rather inappropriate. The Stealth bomber, though hardly an "aeroplane," is virtually all wing, and some of the contraptions visible at air shows can seem to have but vestigial wings. For all the vagueness of their claims, the Wright brothers, like all inventors, were merely attempting to forestall the inevitable alternative designs and improvements others would make in the flying machine, just as the Wrights had discovered and articulated the shortcomings of the aeroplanes and other components whose elimination made possible the first sustained manned flight. As invaluable and unique as those components might have seemed at the time, in the final analysis they were not what they were cracked up to be. And they certainly were of no unique form.

Though the Wrights are remembered for their singular achievement, there were in fact competing designs for the first successful flying machine. These, however, are no more easily remembered than are the competitors of the Gossamer Condor for the Kremer Prize. They ranged from "craft in the tradition of Leonardo da Vinci's ornithopter with flapping wings to pedal-powered machines with two-man crews," but none of them could execute the mile-long figure-eight course. (And it is hard to imagine that many if any of the unsuccessful craft continued to be developed after the prize was won.) Before a notable achievement, technological or otherwise, there is often a goal but no real standard against which to judge competing plans or designs for achieving the goal. Once the goal has been reached, however, the form or formula by which that is accomplished becomes the standard against which subsequent attempts necessarily compete and must be judged. It is little wonder that the form of artifacts then tends to evolve within the rather vague but narrow confines set by patent claims and counterclaims.

Like performance competitions, design competitions make evident the arbitrariness of form, but our awareness of it is often only short-lived. When a group of planners decided that there would be a Great Exhibition of the Works of Industry of All Nations—the first world's fair—in London in 1851, they announced an open competition for the design of a temporary structure to house the anticipated sixteen acres of international exhibits under one roof in Hyde Park. A total of 245 diverse entries was received, but the building committee judged none of them to be suitable and came up with an impractical hodgepodge of its own. Only after it was announced to general ridicule did Joseph Paxton, a gardener and designer of greenhouses, submit his own radical design to the committee and leak it to the *Illustrated London News.* It proved to be the design that was adopted, and the highly successful Crystal Palace became the paragon of exhibition buildings for decades of subsequent world's fairs.

At the close of the Great Exhibition there was another competition—for ideas to reuse the cast iron and glass from the Crystal Palace—and one of the entries proposed building a thousand-foot-high crystal tower. Thus the same modular parts could be imagined to be formed into the tall and narrow as easily as the short and squat, much as the parts of a child's Tinker Toy might become a bridge or a crane. In the twentieth century, the entries in design competitions for skyscrapers have proved time and again that no single form follows the function laid out in the call for submissions. The Tribune Tower in Chicago was the result of a design competition, which included entries that ranged from a skyscraper in the whimsical form of a colossal classical column to the serious Gothic tower that was chosen and built. A recent television documentary tracing the history of a restricted competition for the design of a new central library building for Chicago showed just how diverse proposed solutions can be, and also how function can become forgotten among the considerations of aesthetics, symbolism, and politics that can so influence the final choice.

The Sydney Opera House is a classic case history of design competition and of what can go wrong with large projects. A total of 223 entries were received for the performing-arts complex to be built in Sydney Harbour, and the competition was won by the freehand sketches of the Danish architect Jørn Utzon. His design was a striking assemblage of huge shells that evoked sailing boats, but it omitted any consideration of engineering factors, which made the design highly impractical and the structure extremely difficult to build. Even though the Opera House generally was regarded as an architectural and engineering masterpiece upon its completion in 1973, it opened nine years late and more than 1,400 percent over its original budget. The architect's obsession with (arbitrary) form had necessitated many ad-hoc engineering decisions in the course of the building's erection, and little thought was given to maintenance. In 1989, with hundreds of repair projects deferred and with increasing leaks developing in the opera-house complex, a ten-year rehabilitation program was announced at a cost of $75 million. The form remains one of Sydney's most striking and recognizable visual images, but its function leaves some things to be desired. Unfortunately, the form of an opera house can not respond to failure so quickly as can the form of motorcycles, tractors, or even silverware.

There is one class of large and very visible structures in which form follows engineering, rather than dictating it, but still no single form follows from a stated function. Large bridges constitute perhaps the most pure of engineering structures, and their form is often an expression of the very mechanical principles by which they work.

Some of the most beautiful bridges in the world have emerged from design competitions, and the procedure has been especially effective in Europe. There, competitions not only encouraged but also gave opportunities to such pioneering engineers as Robert Maillart and Eugène Freyssinet to develop new forms along with new construction techniques for concrete bridges. Their legacy is a landscape in which technology and nature harmonize rather than clash.

David Billington, who has written thoughtfully on aesthetics and bridge engineering, believes that design competitions can provide an opportunity for constructive interaction between the general public and the public agencies commissioning designs, and that such interaction can lead to better civic structures. Indeed, according to Billington, public involvement in the design process can have wide-ranging benefits:

> It is relatively easy to take one project and judge it as good or bad; it is quite another thing to take several carefully thought-out designs for the same site, rank them, and then justify that ranking in regard to the concept, details, cost, and appearance. This exercise tests the jury as much as the contestants and forces the jury to explain all aspects of bridge design to the public in a clear and jargon-free report.

Whether with bridges, skyscrapers, or any other structures or machines, it is the initial specification of function that certainly defines the problem to be solved and constrains the solution. But the formulation of a design problem in no way dictates its solution, as the variety of entrants in any competition demonstrates. The requirement of a bridge across a strait or a ravine has historically elicited designs ranging from arch to suspension structures; these may be said to be at opposite ends of the structural spectrum, with the former working in compression and the latter in tension. Which solution was favored by one designer and which by another may have depended as much on a preference for materials (say, wrought versus cast iron, or steel versus concrete) as for construction technique (building from the top down or the bottom up). Political constraints, such as nineteenth-century Britain's headway requirements for the high-masted water traffic that could not be obstructed by an arch, or twentieth-century New Mexico's preference for the flatness of a plateau that could not be punctuated by towers rising above a gorge, might have as much to do with a choice of form as the expected traffic volume influences the number of lanes. Though material, constructional, and aesthetic considerations are arguably as functional as traffic constraints, the nonuniqueness of the ways in which the former can be collectively satisfied or compromised is just one more argument against form's following function.

Design competitions, whether sternly and publicly judged or jovially and privately played out in the office of a commissioned designer, can be counted upon to produce more forms than the functions that drive them. Freedom in the earliest, conceptual design stages can be great fun for all concerned, but it is the serious choices among forms and details that make the difference between ultimate success and failure.

11

Closure Before Opening

A prize of 12,000 francs for a method of preserving foods was announced in 1795 but stood unclaimed for fourteen years. Finally a Parisian named Nicolas Appert demonstrated his scheme for putting cooked fruit, vegetables, and meat in bottles and then immersing them long enough in boiling water to destroy the bacteria that had frustrated prior attempts at food preservation. He set forth the method in his 1810 treatise, *L'Art de Conserver,* which was soon translated into several languages, including English.

Even though they were airtight, bottles were breakable, of course, and this was a distinct disadvantage in transporting preserved food through the heat of battle that soldiers encountered or over the rough terrain that explorers covered. In 1810 Peter Durand, a London merchant, eliminated this shortcoming by employing a "tin canister" for preserving food. The firm of Donkin and Hall set up a "preservatory" in London, and the new tin-coated wrought-iron cans promised to be an excellent means of supplying British soldiers and the Royal Navy with homestyle food away from home. Unfortunately, early efforts apparently focused so much on the objective (or function) of preserving food against spoilage that little thought seems to have been given as to how the food would be removed from the tin. It is the rare artifact that does not require also an infrastructure of auxiliary artifacts to be developed.

The complications necessarily associated with preserving food were evidently the most immediate problems to be faced by the inventor, but having the food preserved so that it could be eaten at will (and away from a blacksmith's shop) was clearly an ultimate function of the tin can. Nevertheless, the preservatory objective so dominated the can's early development that soldiers reportedly had to attack their canned rations with knives, bayonets, and even rifle fire, as American Civil War soldiers still would a half-century later. If Donkin and Hall wanted to sell their products to a broader clientele, they certainly had to address the problem of how to get what was inside a can out civilly, but as late as 1824 a tin of roast veal carried on one of the explorer William Edward Parry's Arctic expeditions bore these instructions for opening: "Cut round on the top with a chisel and hammer."

In spite of this shortcoming of the iron container, English shops were selling canned food to the public by 1830, and the Englishman William Underwood, who established America's first cannery in the early 1920s, apparently spoke for all his contemporaries when he advised using whatever tools might be available around the house to open tin cans in any makeshift way. Despite the need for it, no specialized tool was to be forthcoming for quite some time. In the meantime, early cans made of heavy-gauge iron were "sometimes heavier than the food they contained." For example, the can containing the veal taken to the Arctic weighed over one pound empty, and had a wall one-fifth-inch thick. There soon developed alternatives to hammer and chisel for those not carrying their canned goods on faraway expeditions, however, and "the first can openers may have been elaborate mechanisms with which the shopkeeper opened each can before it was taken away."

The early can was a success at preserving food, but its tolerable failures were obviously its weight, which directly affected its cost, and the difficulty of getting at the food. Having the shopkeeper open a can at the point of sale meant that its contents had soon to be consumed, thus removing any advantage of having the food preserved at the ready in one's own pantry. Such an objection to an otherwise wonderful product was what drove some inventors to concentrate on ways to make cans thinner, lighter, and easily assembled and disassembled, while others tackled the problem of developing specialized tools for opening cans. The replacement of iron with the stronger steel in the late 1850s did make cans thinner, but the greater flexibility of the lighter material in turn necessitated the introduction of a rim for stiffening, and for attaching a top and a bottom, which in earlier times had been folded over onto the sturdy side of the can. (Today, many steel cans are corrugated under their paper labels to stiffen their thin sides further against being dented during handling.)

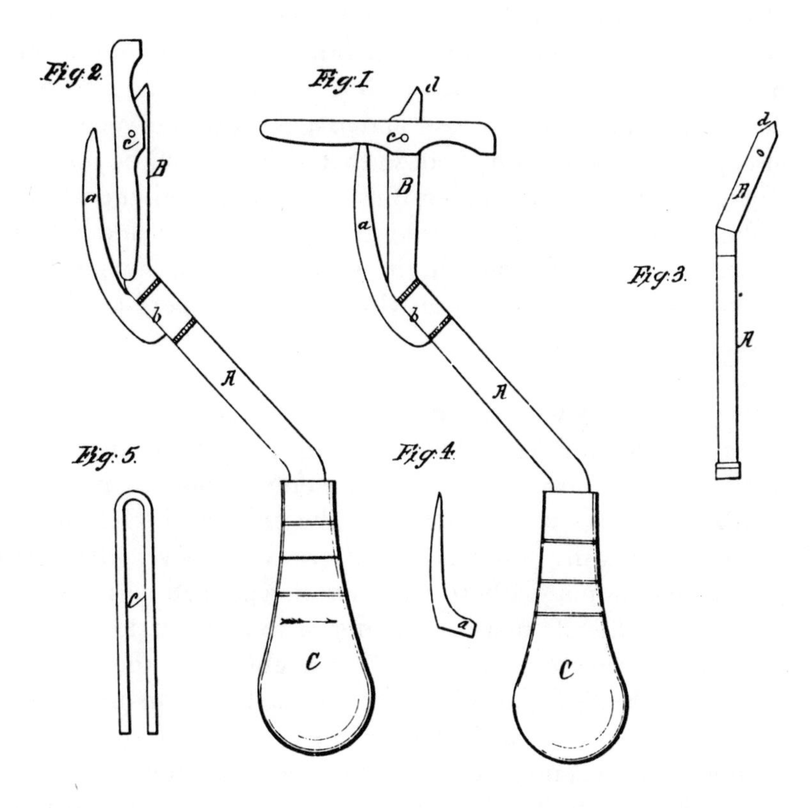

Ezra Warner's 1858 patent for a can opener removed flaws of earlier devices, which often caused the liquid to fly out when the can was first punctured with a percussive blow of a pointed object. With Warner's invention, the can top was pierced not by percussion but by pressing in the point *d*, which was prevented from penetrating too far into the can by the guard *c*. With the top breached, the guard was swung out of the way so that the cutter blade *a* could be worked around the can top.

In 1858 Ezra Warner, of Waterbury, Connecticut, obtained a landmark patent for a can opener. Described by one student of the origins of everyday things as "part bayonet, part sickle," the large curved blade had to be worked forcibly around the can's periphery. Like inventors before and after him, Warner defended the form of his brainchild by comparing it with more primitive forms and implicitly pointing out their shortcomings and outright failures:

> The advantages of my improvement over all other instruments for this purpose consist in the smoothness and rapidity of the cut, as well as the ease with which it is worked, as a child may use it without difficulty, or risk, and in making the curved cutter susceptible of being removed, so that if one should be injured it may be replaced by another, thus saving all the other portions of the instrument, and consequently much expense, and in that the piercer will perforate the tin without causing the liquid to fly out, as it does in all those which make the perforation by percussion of any kind.

Although such devices were in some use during the Civil War, soldiers and homemakers alike had long become accustomed to making do with more familiar implements to open cans, and so specialized openers were not necessarily employed. Only in 1885 did the British Army and Navy Co-operative Society, whose catalogue was an omnium-gatherum of Victorian gadgets and merchandise, seem to offer its first can opener. The 1907 catalogue of the cooperative offered several "knives" for opening tins, including one known as the Bull's Head. Thought by some to be the first popular domestic can opener, it had a red handle cast in the form of a bull's head at its working end, while the other end had a bull's tail looping nicely back upon itself to form a graceful handle. A screw through the bull's neck held an L-shaped blade that formed the animal's lower jaw and provided the cutting edge of the opener, which, like virtually all of its kind, worked on the principles of the wedge and the lever. The other end of the blade projected out from the bull's withers, and was no doubt convenient for piercing the top of the can as a first step in opening it without bending or breaking the necessarily longer cutting end of the blade.

Anyone who has used an old-fashioned can opener, whether or not its form suggested a powerful animal, knows all the disadvantages of the tool. Its action is jerky rather than continuous, and the jagged edge left behind has been the cause of many a cut finger. The first opener with a wheel for cutting in a more continuous and smooth fashion was apparently patented in America in 1870 by William Lyman of West Meriden, Connecticut. One end of his opener was used to pierce the center of a can's top and serve as a pivot about which the opener's handle pulled a cutting wheel. The device had

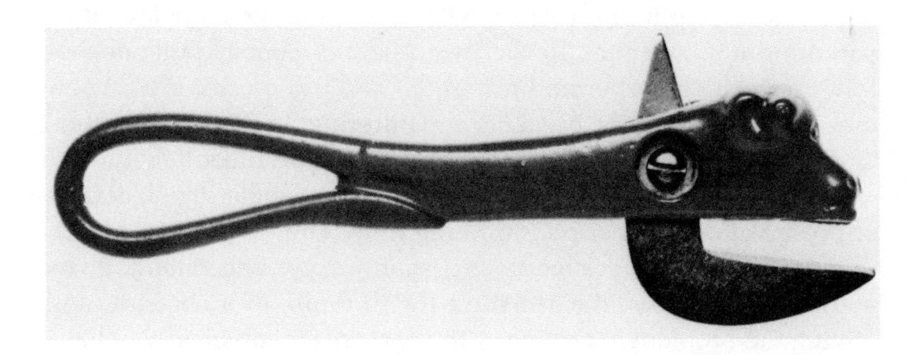

The Bull's Head can opener consisted of a cast-iron frame whose head gave the object its name and whose handle continued the fanciful theme. The L-shaped blade, pivoted about a screw, had one short and sharply tapered end for piercing the can's top without penetrating so deep that removal was difficult. The longer end of the blade worked in a familiar manner. As can tops became thinner and easier to pierce, a single blade could serve both functions.

to be adjusted for each size of can, and its efficient operation relied upon getting a bull's-eye with the piercer.

In 1925 a patent was issued for an improvement on what has become a more familiar style of wheeled opener, one that pinched and rode around the rim of the can. This improvement employed a serrated wheel to reduce slipping. The 1928–29 Sears, Roebuck catalogue offered an "up to date can opener" called the Simplex, which had a serrated gripping wheel and a cutting wheel that worked around the side of the can to remove the "entire top," including the rim. Now, of course, there is a great variety of can openers, including electrically operated ones, but each of them has its own shortcomings, drawbacks, inconveniences, or little annoyances. Those that work with a squeeze of the handle and a twist of the wrist can be tiring to use on large cans, and frustrating when their driving wheel slips and fails to grip the can. Electric can openers, on the other hand, can be bulky counter-cluttering devices that are difficult to clean. Almost two centuries after the introduction of tin cans, there is still room for improvement in what might be termed the infrastructural equipment to break into them to remove their contents, and thus there are likely to continue to be inventors patenting new openers. In the meantime, of course, pull-opening tops are being incorporated into more and more cans, giving new meaning to the phrase "breaking and entering," and so the question of developing a better can opener may become moot.

The generic problem of meeting the often conflicting objectives of preservation and access is nothing new. The frustrations of getting at what nature packaged was certainly experienced long ago by many a tropical islander thirsting for the milk of a coconut, and solving the problem of getting at the contents seems clearly to have been more the consumer's than the packager's problem. Perhaps one of the most culture-laden of artificial beverage containers is the wine bottle, which has such strong traditions associated with it that even the slightest variations in form or color have come to be associated with different wines. It might easily be argued that the present form of certain wine bottles has from the first followed their function, but such reasoning would likely be after the fact. For example, the characteristics of champagne bottles—their heavy thickness, their punted bottoms, and their thick lips that provide an anchoring device for their mushroom-shaped corks—are all well suited to containing highly pressurized champagne while at the same time minimizing breaking, exploding, spontaneous uncorking, or the need for a corkscrew. It is less likely that all these characteristics were present *de novo* in champagne bottles than that they evolved one by one as the more conventional bottles in which champagne was first stored broke, exploded, or prematurely and unceremoniously popped their corks.

The different shapes of bottles in which, say, Rhine and Burgundy wines are stored more likely have their origins in accidental local variations and evolutionary changes in bottle making than in any prescribed subtle functional advantages of a long or a squat neck. Though it is possible to argue the advantages of one neck over the other in reducing the sediment decanted with the wine, say, it is most probable that that feature, if not just a happy accident of place, developed when decanting sediment with red wine became an unacceptable annoyance to at least one inventive mind in a position to do something about it. Thus, the functional correctness of putting sediment-prone red wines in bottles whose shoulders can trap sediment is more likely a result of the ruining of many a glass of wine decanted from earlier containers than the result of the anticipatory planning of some omniscient vintner. Conversely, putting sediment-free white wines in step-necked bottles would have required them to be upended to drain the wine. How much more elegantly emptied is the bottle with the long tapering neck.

The importance of bottle shape was underscored in a recent dispute between the government and a maker of a fortified wine called Cisco. The bottling of this potent wine, containing 20 percent alcohol, made it look like a wine cooler, whose alcohol content is only about 4 percent. Because of the similarity in packaging, stores shelved Cisco with wine coolers, and the more powerful drink was reportedly linked to alcohol overdosing and violence among teenagers, who came to call the new stuff "liquid crack." To avoid future confusion between its fortified wine and the lighter coolers, the manufacturer declared that it would put Cisco in a newly designed bottle, one that would be "mature and masculine; certainly . . . unlike any wine cooler on the market."

Even the color of wine bottles can be attributed to evolution fixed by tradition rather than by any firm functional determinism. Green and brown bottles are more likely to have evolved after sunlight was recognized to ruin wine in clear bottles than to have been devised in anticipation of the failure. But even invoking this argument is not to say that changes in form must follow recognitions of failure, for, though Sauternes are probably also affected by sunlight, they have traditionally been bottled in clear glass.

Regardless of its shape and color, a wine bottle must be sealed to protect its contents, and the cork is a natural sealing device. But, as effective as the cork is in helping the bottle perform the function of preserving wine, it is also a nuisance when one wants finally to open the bottle. Not only can the wine be ruined by a moldy cork, contaminated by a crumbly one, or made inaccessible by a stubborn one, but also we need an ancillary device to remove even the most accommodating of corks from unpressurized bottles. (The pressurized champagnes no doubt inspired the mushroom-shaped corks that can be coaxed out with the thumb after many a corkscrew-wielding hand had been wrenched back by a cork missile.) Like can openers, corkscrews and related devices have proliferated as the shortcomings of each existing one gave rise to a new, improved model. A few are almost foolproof tools, but even the most reliable can fail when encountering a bad cork. Some wine makers will confide *sotto voce* that real corks are an unnecessary expense and risk in these days of plastic, and that even the glass bottle itself is an unnecessarily awkward and expensive container for wine, but tradition is a strong persuader, especially in the wine industry, and only the least expensive wines tend to be sold in bottles with screw tops or in boxed bags with convenient spigots.

The bottling of beers has its own traditions and prejudices, of course, and they can seem to be as sacrosanct as those of wines, but uncapping a bottle involves an action different from extracting a cork. Yet, as if to acknowledge their roots, it was not so long ago that metal bottle caps had cork inserts, which were pulled tight against the mouth by the crimping action of the cap around the lip on the bottle's neck. This was a relatively easy action to mechanize, but it also required a unique maneuver to undo the cap in order to drink the contents. When I have found myself with a bottle of beer and no opener, I have realized how difficult it can be to get the cap off without the specialized tool that did not exist before the cap. I have never been thirsty or brave enough to resort to using my teeth, but I have been able to find makeshift openers in the various nooks and crannies of door hinges and drawer pulls. It is also effective, even if time-consuming, to loosen with a nail file or a fork tine each of the crimps in succession around the cap until it can be pushed off with the thumb. What is common to all these emergency actions is that they rely on the mechanical principle of the lever; indeed, virtually all bottle openers have continued to work on that same principle.

As the development of the can opener followed at some distance the development of the tin can, so the specialized bottle opener emerged only after the bottle cap itself. As with cans, there is clear evidence that opening a bottle was not given as much thought as sealing it. In the early years of this century, for example, many more patents for bottle caps and capping machines appeared before patents for bottle openers did, and over the first decade of the 1900s patents for bottle-capping devices outnumbered those for opening them by about ten to one. Certainly the more immediate objective of bottlers was to keep their beverages fresh and intact in transit to the consumer, but how the customer was to open a bottle of beer should also have been a consideration in design and commerce.

The inconvenience of requiring a special opener to uncap a bottle is what led to the development of the screw top that is so familiar on beer bottles today. But, again, tradition and prejudice can affect whether a new form, even one that is a clear improvement in technology or a clear advantage in use, will be universally adopted. One of the disadvantages of requiring openers was that beer companies often had to supply them free, as matches were with cigarettes, lest the consumer be frustrated in attempting to consume the product. If the need for openers were eliminated, the expense of selling beer could be reduced—a clear advantage. This cost savings would naturally be most significant for lower-priced beers, which tended to have the largest volume of sales, and so these brands were more likely at first to embrace the new technology. This in turn meant that lesser-quality brews were associated with the twist-off cap, and so it has been somewhat eschewed by bottlers of premium and imported beers.

Soft drinks were long bottled much the same way as beer, and stationary openers were usually attached to the cooler or machine where the soft drink was purchased. Since, unlike beer, soft drinks tended to be consumed on the spot, this was no great inconvenience. However, another disadvantage of bottles has dominated the evolution of beverage containers: the logistics and cost of collecting and refilling them. When bottles were expected to be reused, they had to be strong and tough enough not only to hold their contents but also to survive the abuses of repeated handling, transportation, and washing by both humans and machines. Since chips, nicks, and scratches weaken a glass bottle just as they do a windowpane, it was necessary to make early bottles especially heavy. The twenty-four-ounce-capacity bottles that Montgomery Ward sold for home use in 1922, for example, weighed almost two pounds each.

A beverage container that was disposable like a tin can would be a much better solution from the beer or soda company's point of view, of course—if the customer would accept the idea and pay for it. Neither consumers nor merchants would have to devote space to collecting empty bottles, and there would be transportation and sanitary advantages. The plastic soda bottle invented by Nathaniel Wyeth was one way of dealing with the objections to glass bottles for soft drinks, and the features of the screw-top plastic bottle have clearly developed in response to the failures of its crimp-capped glass counterpart: removing the inconvenience of needing a bottle opener, reducing the weight to be carried to and from the store, and eliminating problems associated with breakage and germs. Unfortunately, as is not infrequently the case when the evolutionary process takes place at revolutionary speed, the newer technology is not without its own shortcomings and disadvantages. Because they are lighter, plastic bottles can be made in larger than traditional capacities, which in turn keeps their unit cost down. But larger bottles can be unwieldy to pour from, and the soda often goes flat well before the plastic bottle is empty. However, the problem of what to do with used plastic bottles may be their greatest single shortcoming at the present time, as it is with virtually every single-use type of container or packaging.

The disposable can has developed into another alternative to the glass beer or soda bottle, but at first beverage cans were not much different from tin cans for food. In particular, they were formed from three pieces of tin-plated steel: one rectangular piece bent into a hollow cylinder and welded along its seam, with two circular disks for the top and bottom. And, of course, the can required an opener, but, since the contents were liquid, only a hole large enough to make a pouring spout was required. Indeed, anyone who tried to open a can of beer by jerking a Bull's Head opener around its rim would have sloshed the contents all over the place, not to mention risking a jagged lip of first steel and then flesh. Thus, the specialized beverage-can opener known as a church key was developed to pierce the pressurized can with minimum jolting and make a wedge-shaped opening. Ideally, a single pie-slice wedge—i.e., one that extended to the center of the can top—would have allowed the can to be opened with only one motion, and the long opening would have enabled air to enter the can as its liquid exited. However, because early beer cans had relatively heavy steel tops, the applied mechanics of the opener played a role in determining its form, which dictated that it make a much smaller wedge-shaped incision close to the edge of the can.

A church key is a simple lever whose fulcrum hooks under the top lip of a can. The handle extending outward from the can provides one arm of the lever, and the pointed cutting edge extending over the can top provides the other. As with all levers, the length of the handle magnifies the effect of the force applied to its end, but, by the same token, the piercing force diminishes as the distance from the fulcrum to the tip of the cutting edge is increased. Thus, in order to make a church key that is not too long (cost is proportional to amount of material used) and yet is capable of piercing the can top without being bent out of shape, a compromise opener was developed that produced a relatively small hole close to the edge of the can. Drinking beer through such a hole is only slightly less objectionable than drinking it through a straw, and pouring it is a slow, gurgling procedure. Therefore, a venting hole on the opposite side of the top came customarily to be made. (Homemakers were used to putting two holes in a can top, for condensed milk had long come in tin cans that were opened by stabbing the top at two points with the tip of an old-style can opener.)

Specialized tin cans were precursors of what was to replace the steel beverage can. Sardines were always a problem food to pack and unpack, for they were to be served whole and yet they flaked and fell apart easily if poked with a fork or caught on the ragged edge of the can. Because sardines are so fragile, they came to be packed in tins that allowed the can to be laid flat. Furthermore, since a conventional opener would slash the contents of the can before exposing them, a special key was soldered to the bottom of the tin so that its top could be opened cleanly and completely by being rolled back upon the key, thus presenting the tightly packed fish as whole as they could be. To this day, special sardine forks sold by the German silversmith Wilkens, for example, have widely spaced tines to give plenty of support to the sardine, lest it break while being lifted, and the points of their tines are connected by a silver bar so that they cannot pierce and flake the fish in serving it.

The idea of a sardine can long survived in such diverse applications as cans for coffee, peanuts, and tennis balls. These cans no longer come with keys attached to their bottoms but, rather, have pull rings riveted to their tops, which are scored along their periphery, where they are designed to fissure, and indented across their width to give them sufficient stiffness so they will not buckle and pull the sides of the can together in the process of opening. With the proper design of fracture lines and stiffening ridges, a top can be removed in a predetermined way without a separate opener and leave no rough edges to scratch the contents or the hand reaching for them.

Some consumers seem less squeamish than others about the use of cans. There is a television commercial in which a big burly guy crushes a beer can against his forehead, and I get a headache every time I see it. Even though I know that today's beer cans are pretty flimsy, and that squeezing the sides of the can just as it strikes the forehead makes it collapse harmlessly, my childhood memories of tin cans overrule any adult understanding I might have. I have yet to summon the courage to test my engineering predictions by crushing a can against my own forehead.

A good deal of our visceral sense about how physical things behave is formed in our childhood, when we have more time and fewer inhibitions about looking closely at and experimenting with the stuff that we find all about us. My own sense of the strength of a beverage can was probably established by the time I was about seven years old. That was in the days before television occupied children's afternoons, and my friends and I looked for entertainment wherever we found it. Coming upon an empty can in the street could keep us busy till dark.

Whoever among us found the can would stomp on its side until the top and bottom curled around his shoe and locked into place like the clamps on an old roller skate. The can fit our foot like a clodhopper and, as we walked along the concrete sidewalk, made a noise heard round the block. As our group came across other empty cans, we would stomp them into more tin overshoes and have a grand time making noise and seeing who could wear the cans as shoes longest.

Getting a good fit with a tin can was no simple matter, for the cans seemed very strong to the foot of a seven-year-old, and a misdirected stomp that hit the unyielding end rather than the side of the can could be felt for days. At the same time, once the top and bottom had begun to curl around the foot, a more delicate touch was required lest the makeshift overshoe fit too tightly. Stomping cans in hard-soled shoes worked best, but we often wore canvas sneakers—high-topped Keds—and in those our feet were especially vulnerable to the revenge of the heavy tin can, if we could get the noisy toy to hold to them at all.

After such childhood experiences with it, the can as beverage container held little interest for me when I grew older. I have certainly bought my share of six-packs, but the cans themselves were not the focus of my attention. I thought that a can was a can—unless it was to be made into a kid's shoe. But we were not kids anymore, and none of my college buddies ever even joked about smashing a can against his forehead. If we had been asked what we thought would happen if we did such a thing, we would probably have said something between a large gash and a frontal lobotomy.

As the television commercial demonstrates, the evolution of beverage cans has outstripped previous generations' understanding of them. What had happened, while my friends and I had been growing into middle age, to turn the head-gashing instrument of the 1950s into the collapsible cream puff of the 1990s? Like all technological change, the story of the beverage can involves considerable interplay between engineering and social factors, not the least of which are economics and the environment.

In the late 1950s, I was aware of few complaints about beverage cans. In fact, they were convenient but otherwise unremarkable things, although there may have been some talk about a growing litter problem. Aside from their taller shape, beer cans were not unlike the familiar tins containing food, but opened with a church key instead of a can opener. However, while consumers drank contentedly, the brewing industry was concerned about the steadily rising cost of tinplate—the tin-coated steel out of which the cans were made. Kaiser Aluminum had initiated research-and-development efforts in the early 1950s and produced a lightweight and economical aluminum can in 1958. At the same time, the Adolph Coors Company and Beatrice Foods had joined in their own research-and-development program, and in early 1959 the first Coors beer was sold in seven-ounce returnable aluminum cans the brewer made itself. (Hamm's and Budweiser did not get their first lightweight cans for another four years, when they were able to buy them from Reynolds Metals and Alcoa, respectively.)

The new cans were revolutionary not only in their raw material but also in how they were made. Whereas the relatively heavy old tin cans comprised three pieces, an aluminum can begins with a disk of metal that is first pushed into the shape of a cup that looks not unlike a tuna can, and then it is stretched to make the taller sides of the one-piece bottom. After the can is filled, a top is crimped on. This same basic procedure is used to make today's aluminum can, though various improvements have been incorporated over the last three decades, especially in reducing the amount of metal involved. In the early years, one pound of aluminum made fewer than twenty cans; today, almost thirty cans come out of the same amount. The thickness of the can wall is less than five-thousandths of an inch, about the same as a magazine cover.

The walls of the can can be so thin because its contents are under pressure. Just as a flabby balloon stiffens when it is blown up, so the

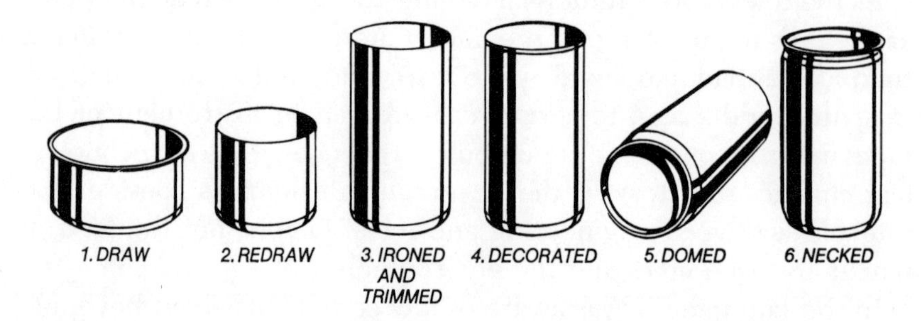

There are several steps in forming a seamless aluminum beverage can: (1) a flat circle of metal is punched into a tuna-can shape; (2) it is drawn out to a taller shape; (3) it is squeezed out to its final height; (4) it is printed to advertise its contents; (5) its bottom is given a characteristic domed shape to resist the pressure it will contain; and (6) its neck is formed to be crimped around the top that will be added after filling.

carbonation in a beverage can stiffens it. However, a flat can bottom would also round out like a balloon and make the can rock on the store shelf or kitchen table, and so the characteristic inward dishing of the bottom of an aluminum can is necessary. Because a convex face is put against the pressure, the bottom acts like an arch dam in resisting the pressure of the fluid behind it, much the way the punt does on a champagne bottle. The can top, on the other hand, cannot be so dished, and thus it must be thicker than the rest of the container. (To save metal in the thicker top, aluminum cans have come to have the characteristic stepped neck, which requires a smaller-diameter top: reducing the diameter of the top as little as a quarter of an inch can save 20 percent of the metal required to make it.)

Although the tops on the first aluminum cans were noticeably easier to open than the steel variety, a separate opener was still required. This remained a clear disadvantage, especially when there was plenty of beer but no church key at the family picnic. It was in such a situation that Ermal Fraze of Dayton, Ohio, found himself in 1959, when he resorted to using a car bumper to open a can. The operation evidently yielded more foam than refreshment, and Fraze is reported to have said, "There must be a better way." On a subsequent night, after drinking too much coffee, he was unable to sleep and went to his basement workshop to tinker with the idea of attaching an opening lever to a can. He was hoping the activity would soon tire him out so that he could get to sleep, but, according to Fraze, "I was up all night and it came to me—just like that. It was all there. I knew how to do it so it would be commercially feasible." Fraze could make such a judgment because he was the owner of the Dayton Reliable Tool and Manufacturing Company, and he had considerable experience with metal forming and scoring, the mastery of which would be essential to developing the pop-top can, for which he obtained a landmark patent in 1963. "I personally did not invent the easy-open can end," he later asserted. "People have been working on that since 1800. What I did was develop a method of attaching a tab on the can top."

Eventually a ring that functioned as a lever was riveted to a prescored tear strip (whose shape could be reminiscent of that of a Schmoo from the contemporaneous *Li'l Abner* comic strip), and a lever action enabled the ring's tab to break the can's seal. Then a pull on the ring removed the attached strip of metal from the can top in a manner not unlike the way a perforated mail-in coupon is removed from a magazine. Because of the lever action and preferential scoring, the can opened first at the top of the hole, and a further pulling action tore the metal free of the can along the characteristic scored outline. The hole that was left extended a good distance from the edge of the can, to (or beyond) the center, and so, as the can was tipped for pouring or drinking, air could enter the top of the hole and allow the easy, gurgle-free exit of the contents. The early pop-top or pop-tab can worked reasonably well, not only eliminating the need for a church key but also reducing the action of opening a can from making two separate triangular incisions on opposite sides of the top to pulling a single ring in what would ideally be one smooth motion.

Still, scoring a tear strip into a can top so that it will be easy to remove but yet strong enough to hold against the can's pressure requires some rather tricky engineering in the way the metal is formed. Some early pull tabs were being blown off prematurely by the high pressure of the carbonation rushing out of the hole made by a consumer's initial cracking of the tear strip, and so Fraze and other inventors came up with schemes to direct benignly the first whoosh of escaping gas away from the tab itself. Throughout the mid-1960s, numerous patents were awarded for improvements in

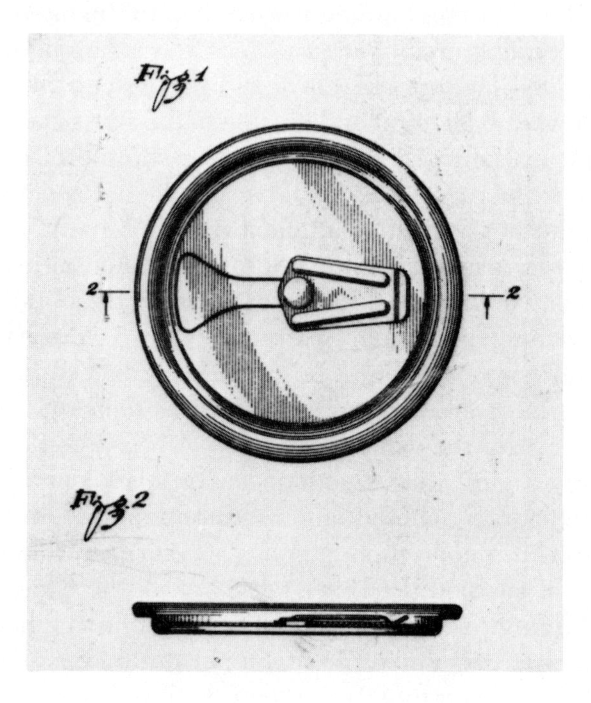

In the early 1960s, Ermal Fraze applied for various patents related to self-opening cans and their manufacture. There were many difficulties to overcome, for it was a touchy matter to make the can easy enough to open and yet keep the tab from pulling loose or the top from opening prematurely. He received a design patent in 1963 for this "ornamental design of a closure with a tear strip opener."

pull-tab devices, but then a new problem arose with them—environmental pollution.

By the mid-1970s, the tabs that were pulled completely off the can top were coming under increasing attack from environmentalists, and with good reason. I recall stopping at traffic lights in those days and trying to count all the pull tabs (looking somewhat like little curled-up tongues on key rings) among the cigarette butts beside the road. I could never finish counting before the light changed. Picnic sites and beaches were especially full of the sharp litter, which was difficult to clean up because the small tabs passed easily through the tines of rakes used by cleaning crews and beachcombers. (One young boy was reported by *The New York Times* as having collected twenty-seven thousand in an attempt to qualify for the *Guinness Book of World Records.*) Animals and fish, not to mention children, were swallowing the tabs, and they were cutting the feet of many a bather. Rather than discard it, some conscientious people would drop the tab in the can after opening it, but some of them required operations when they swallowed the tab with their drink. In short, there was growing concern over the failure of the pull tab to function as well as desired, and this led to another rash of patent applications for easy-open cans without removable tabs.

There were several clever schemes to solve the loose-tab problem, and Coors was again in the forefront. It developed a two-step opening procedure, in which a protruding button of scored metal was first pressed to break the pressure seal. A second, larger button was then pressed into the can to provide a drinking hole. Restoring two-step opening procedures did not prove to be very popular, however, and their shortcomings, which included the relatively large shove required to open the can and the need to push a button through the sharp edges of a hole, were not lost on inventors. They happily included in their patent applications, as a description of the prior art, the failings of existing solutions to provide an "easy open ecology end" for the beverage can. A bewildering number of patents was being issued by the mid-1970s, but many of these were merely variations of the familiar pop tab that prevented it from being pulled all the way off.

In 1975 a patent was issued to Omar Brown of Kettering, Ohio—but the rights were assigned to Ermal Fraze, the inventor whose name seems to be virtually synonymous with easy-open can patents—for a "can end with inseparable tear strip," and in a section giving some background to the invention, an especially vexing problem associated with simply folding the tear strip over the top of the can was noted:

> Since most people drink the contents directly from the can, it is quite probable that the user's nose will contact a tear strip which is not fully removed from the can. If the edge of the strip is sharp, it is possible that he may cut his nose on it. On the other hand, if a sharp edge is formed around the pour spout, he may cut his lips on it.

When removable pull tops came to be recognized as a significant litter problem and safety hazard, can manufacturers searched for alternatives. The Coors firm, which had pioneered the aluminum beer can, came up with an "environmental package." Six cans were sold held together with drops of glue, thus eliminating the need for any other wrapping, and the cans were opened by first depressing a small button to break the pressure seal and then punching in a larger button to provide a drinking hole. This awkward design soon led to today's familiar can top.

Brown's solution to the problem included recessing the pour spout, thus keeping the lips from its sharp edges, and having the opened tear strip lie flat against the can top and away from the drinker's nose. Another Ohio inventor, Francis Silver (who also assigned his patent to Ermal Fraze), protected the drinker by forming the tear strip so that it could be folded up between the can top and the pull tab. No solution proved to be wholly satisfactory, for each had its own apparent failings, not least of which was leaving too much sharp and sticky metal bunched up on top of the open can. The version of the inseparable tear strip that is on almost all beverage cans today appeared around 1980 as a variation on the Coors push button, but operated on the lever principle through an attached tab. Since the tear-strip panel is pushed into but remains attached to the can top, both the litter problem and the danger of swallowing a tab or cutting one's nose on a sharp piece of metal are virtually eliminated.

Before environmental and worse problems with pull tabs became evident, soft-drink companies also began to package their beverages in aluminum cans. Steel cans were never totally satisfactory for soft drinks, because a church key was required to open them, and that was not in the tradition of soda drinkers. When the pull tab removed the need for an opener, the aluminum can first developed for beer was adopted for soft drinks as well. In 1965 Royal Crown (now better known as RC) Cola became the first to use lightweight cans; Coke and Pepsi followed in 1967. Indeed, because the absence of bottom or side seams on the new cans also made it possible to decorate them in much more elaborate ways than the old tin cans, aluminum was enthusiastically enlisted in the Cola Wars. Other advantages of the lightweight can included its lower transportation costs, its compactness, its ability to be stacked more securely, and the fact that it eliminated having to deal with empties.

The one-time use of cans began to argue against them, however. By the early 1970s, beer and beverage cans were being emptied at the rate of thirty billion a year in America, and ban-the-can bills were under consideration by a majority of state legislatures. Tin-plated steel cans, which were still in the majority, would at least rust in landfills, but the increasingly popular easy-open aluminum cans would not. As Coors seems to have recognized from the beginning, recycling aluminum cans was not only the environmentally responsible thing to do, it was also essential to the long-term acceptance of the new technology.

When the disposal of cans came under the growing scrutiny of environmentalists and lawmakers alike, the industry began to keep records on recycling. By 1975 about one in four aluminum cans was being recovered, and by 1990 the rate exceeded 60 percent. It is the joint goal of the Aluminum Association, the Can Manufacturers Institute, and the Institute of Scrap Recycling Industries to have a reclamation rate of 75 percent by 1995. This not only makes sense environmentally, but is also good business. Recycled cans are essential to supplement the general aluminum supply, and the collection infrastructure is now so efficient that the metal in a used can may reappear in a new one in as few as six weeks.

By 1990 aluminum accounted for about 97 percent of all beer and soda cans made in America, and about 70 percent of all U.S. beer and 50 percent of U.S. soda was packaged in them. In contrast, about 95 percent of all food cans (about thirty billion per year) remained tin-plated steel, because an economical aluminum container is not strong enough to keep its shape without the pressure of carbonation. We may begin to see more aluminum food cans in the future, however; the industry is developing strengthening techniques that include injecting liquid nitrogen into food cans to provide pressure and corrugating the can walls to provide dent resistance.

To counter its failings, the steel-can industry is engaging in research and development of its own. The economics have worked against steel beverage cans in part because they have had to be fitted with aluminum tops to be easy-opening. Even though steel has the recycling advantage of being magnetically separable, the presence of the aluminum complicates the recovery of the metals. A new ring-pull can end made of tin-plated steel may remove that objection, if it can be made as easy to open as an aluminum one and if its exposed edges can be made as smooth. The Steel Can Recycling Institute was formed in 1988 to promote the recycling of tin-plated steel cans, and it hopes that recovering food cans will protect its sponsoring industry. As an alternative strategy, steel-can manufacturers are developing plastic cans suitable for microwaves.

Although perhaps a trillion aluminum cans have been produced and their contents consumed over recent decades, and even though hundreds if not thousands of patents have been issued for improvements, the form has not necessarily been perfected. The opening in the newest pop tops is generally oval-shaped and does not extend completely to the edge of the can or toward the middle, where the ring is attached. Hence, pouring and drinking are a bit tricky: tipping a full can too severely does not allow air to enter so readily, and a nearly empty can must be tipped almost upside down to get at the last of the beverage, so it is nearly impossible to empty the can completely. We tend to adapt to available technology, however, and we seem to have come to tip our cans in the same way we came to tip our bottles, at just the right angle for the level of the contents. However, unlike bottles, whose narrow necks gave us plenty of room to maneuver, the tabs attached to pop-top cans do come up to meet our noses if we are not careful. They no longer pose a threat of cutting us, but they do restrict how far the container can be tilted, and so we must compensate with a greater angle in our necks.

But the interest of inventors is not limited to anatomical inconveniences. Among the functional imperfections of the familiar beverage can are its inability to be reclosed if its contents are not all drunk at once. Cans of coffee, nuts, and even tennis balls typically come with plastic lids that can be used to reseal the opened container, but beverage cans generally do not. Although those who sell beer and soft drinks, or even those who consume them, might not see this as a particular disadvantage, such are indeed the failings that have attracted not a few inventors. One was Robert Wells of Steamboat Village, Colorado, who in 1987 was issued a U.S. patent for a reclosable self-opening can end. In presenting the background to his invention, he summarized what he saw as the failings of existing beverage cans:

> While container reclosure may be relatively straightforward with bottles using screw-on caps, reclosing the typical beverage can is another matter. The tear-out panel associated with the typical easy-open can generally is deformed and/or positioned within the can below the end wall during the opening procedure, and thus is unavailable to reclose the opening in that wall. Prior-art expedients to overcome this problem generally have utilized separate stoppers, purchased as accessories, intended to fit on the end of an opened can and temporarily plug the opening. These separate stoppers are relatively small and easily misplaced or simply forgotten, and thus are usually unavailable to someone wanting to reclose an open beverage container. Fur-

> thermore, the structural variations between easy-open ends supplied by different manufacturers make it difficult to provide an accessory stopper which effectively works with the variety of beverage cans commonly available to consumers.

Wells's patent, like most associated with the deceptively simple can-opening (and reclosing) schemes, is a rather long one, with fifteen claims and forty-seven drawings showing variations on his ideas for a device that can be slid or rotated into position to plug the opening in a can top. Many of his ideas seem much too complicated to be practical for a twelve-ounce beverage container, and it is unlikely that a reclosing device will become standard on the pop-top can, even though Wells has made a wholly rational case for the need for such an improvement. Many of us have no doubt experienced an opened can of beer or soda going flat, but we might prefer to drink our drinks more quickly, or even discard their flat remains, rather than deal with a touchy device to reseal them. People seem generally willing and able to respond to the imperfect artifact by modifying their use of it, not complicating it. Inventors, on the other hand, by concentrating on how to correct the failings of artifacts, appear willing to complicate, at least in their initial attempts to remove shortcomings. If the complications are adopted, they become the subsequent challenges to consumers to use and to other inventors to simplify.

Another shortcoming of pop-top cans is the close-fitting pull ring or tab. It is difficult for people with arthritic fingers to get them under the pivoting device to bend it up and break the seal on the can. In order not to risk broken fingernails, the user may have to get out a pen or pencil and wedge it under the can lever to lift it to where it can be grabbed. It is no surprise that two California inventors, Robert DeMars and Spencer Mackay, have recognized this shortcoming. In 1990 they received a patent for their beverage-container opening-and-resealing device, which they justify first for its ability to reseal cans, thus keeping the beverages in opened containers from going flat and thereby conserving the energy that went into producing them. DeMars and Mackay acknowledge that there exist inventions to reseal cans, but they point out that the devices "have not achieved any significant market acceptance." They continue to set the stage for defending the advantages of their own device by explaining:

> It is believed that the reasons that market acceptance has been lacking is that the devices are complicated and inherently expensive, therefore, significantly increasing the cost of the beverage container to the consumer. Also, such devices are somewhat complicated to operate and at times may be difficult to operate by older people or people with arthritis or other afflictions.

The device of these inventors owes its novelty principally to a little hill, or a "camming protuberance," that juts up from the can top. To open the can, the tab is rotated onto this hill, thus lifting up one end. This action not only pushes the other end of the tab into the scored can opening, thus breaking the seal and beginning the opening, but also lifts the end of the tab sufficiently above the top of the can so that even the stiffest and stubbiest of fingers can get a hold to complete the opening procedure. Closing the opened can is effected by peeling a protective covering off the now exposed bottom of the tab, revealing an adhesive underside that can be folded down, rotated into place, and wedged over the opening by the action of the hill. This procedure takes five figures to explain in the patent, and thus may appear to be as complicated as other resealing schemes. But if the resealing aspect were discarded, a beverage can with something like a "camming protuberance" could be a beautiful sight for sore hands.

Independent inventors will no doubt continue to come up with ingenious schemes to respond to objections to current mechanisms for opening cans, but the companies that make and fill the cans will no doubt continue to focus their attention on their prime objective of preserving the contents in the most effective and competitive way. Of late, technical questions relating to the pros and cons of the availability, formability, and printability of steel versus aluminum have tended to dominate design and use decisions that ultimately have affected the form of beverage cans, and considerations of ultimate convenience and usability to the consumer have tended to be crowded out of the corporate if not the inventor's mind.

Since consumers tend to adapt to the pop tops that are generally available, there is often no business immediacy in exploring or introducing improvements. However, if such improvements would give one brand of beverage an advertising or marketing advantage over its competitors, that would be a change worth considering. On the other hand, there can also be a competitive risk in introducing an innovation that might prove to be too radical a change in form or function, and therefore be eschewed by the public. But, finally, if environmentalist or consumer concerns can be articulated as some kind of failure, as were those about removable pop tops, then there is a clear incentive for manufacturers to look at the end use of their products and containers as well as the immediate objectives of preserving and distributing them to people who must ultimately consume and dispose of them. Although a manufacturer's concerns may sometimes appear arcane or selfish to the consumer, they are in fact no less driven by failure, whether functional or fiscal, than is any other aspect of the design, codesign, and redesign processes through which the forms of even our most familiar artifacts evolve.

12

Big Bucks from Small Change

More than twenty-three hundred years ago, a series of "mechanical problems" and their solutions was compiled. Though classicists frequently attribute the collection to the Peripatetic School rather than to Aristotle himself, the *Mechanica* is usually grouped with the minor works of the famous philosopher but given scant scholarly attention. However, the thirty-five questions of the work show the considerable interest in engineering matters that was current in ancient Greece, as it must be in any civilization that wishes to function with some semblance of physical achievement, comfort, convenience, reliability, and economy. Indeed, the opening sentence of the *Mechanica* shows that the concept of engineering was essentially no different in Aristotle's time from what it is today. While Aristotle began his introduction with the recognition that "remarkable things occur in accordance with nature, the cause of which is unknown," he immediately conceded that "others occur contrary to nature, which are produced by skill for the benefit of mankind." The skill of which Aristotle wrote is what has come to be known as engineering, whose formal definition in the 1828 charter of the British Institution of Civil Engineers was meant to encompass all but military endeavors:

> Civil Engineering is the art of directing the great Sources of Power in Nature for the use and convenience of man. . . .

This definition so remarkably echoed the words of the *Mechanica* that it underscores the fact that, whatever it has been called over the centuries, engineering has been a timeless pursuit of all civilizations. Indeed, the official definition currently employed by the American Society of Civil Engineers reiterates the continuity of purpose:

> Civil engineering is the profession in which a knowledge of the mathematical and physical sciences gained by study, experience, and practice is applied with judgement to develop ways to utilize, economically, the materials and forces of nature for the progressive well-being of mankind in creating, improving and protecting the environment, in providing facilities for community living, industry and transportation, and in providing structures for the use of mankind.

Though the society may be accused of trying to cover all bases with this definition, the roots in Aristotle's notion of using "skill for the benefit of mankind" are unmistakable. And even if, in the turbulent wake of the Industrial Revolution, the engineering profession has fragmented into specialties, the idea of using natural resources and exploiting physical phenomena, whether or not perfectly understood, for the ends of civilization remains the foremost objective of all engineering, whether preceded by civil, electrical, mechanical, or other societal but not essential modifiers. Yet, no matter how it is qualified, no ancient or modern engineering is without economic considerations, which affect greatly the form of artifacts.

Among the questions asked in the *Mechanica* is one of special relevance in a consideration of form in engineered things. It is question 25, and it reads:

> Why do they make beds with the length double the ends, the former being six feet or a little more and the latter three? And why do they not cord them diagonally?

The first part of the question is given short shrift with a "probably they are of those dimensions, that they may fit ordinary bodies." If they had failed to fit, the proportions of single beds would naturally have evolved to the happy medium where they did. But it is really the second part of the question that points to much more interesting and subtle aspects of how forms evolve. Aristotle's answer is as follows:

> They do not cord them diagonally, but from side to side, that the timbers may be less strained; for these are most easily split when they are cleft in a natural direction, and they suffer most strain when pulled in this way. Moreover, since the ropes have to bear the weight, they will be much less strained if the weight is put on the ropes stretched crosswise than diagonally. Also, in this way less rope is expended.

The parts of this answer dealing with strained timbers and ropes are really just assertions, for Aristotle does not elaborate on what is quoted here. This is consistent with the contemporary analytical understanding of forces acting at angles, which had still to be properly articulated. However, as has been the case throughout history, craft and engineering advances could and often did proceed even in the absence of scientific explanations. Certainly it required no theory of beds to invent the concept of a wooden frame drilled with holes through which a rope could be threaded to form a mattress support. The idea was known almost three thousand years ago to Homer, for in *The Odyssey* Ulysses describes, on his return to a skeptical Penelope, how he made the frame of their bridal bed out of the parts of an olive tree and in it drilled holes through which he threaded leather straps. The bed was unique in being anchored in the roots of the olive tree itself, and in holding a sentimental significance for the long-suffering couple. To them, the bed was inviolable. Knowledge of its origin was proof of Ulysses's identity.

More conventional beds, created not by ancient superheroes but by traditional craftsmen, would not have been so unchangeable. The cost, comfort, reliability, and maintenance of the common bed could easily have guided its evolution: if the timbers or the ropes sagged too much or broke, they would be pulled tighter or made heavier. The method of stringing thongs or ropes to form the bed would have evolved in response to questions of efficiency and efficacy, with the craftsman reacting especially to correct weaknesses that brought beds to be repaired or drove customers to rival craftsmen whose beds did not need repair. But, whatever the comparative merits of crosswise and diagonal stringing patterns, what is clear from the *Mechanica* is that economy of material, and labor, was as much an issue in ancient times as it is now. Given that the basic form of the bed is as rooted in ancient craft tradition as Ulysses's olive-wood specimen was rooted in the ground, there can be little doubt that the initial and upkeep costs of artifacts have always been strong determinants in their evolution.

In a recent article on American rope beds, which have survived in use into the present century, two alternate methods of cording them also are discussed. In the one, the rope is threaded through holes, as described by Aristotle; in the other, the rope passes over pegs; but in both methods the ropes load the wood across the grain and so any tendency to split is minimized. Whatever method was used, however, the rope could be expected to slacken with time, and so special bed wrenches were kept handy to tighten it. This process would no doubt have caused bed ropes to break now and then, most likely just as someone was ready for a good night's sleep on a nice firm bed. At times like that, a pegged bed would have been a godsend, since a broken rope could easily be knotted and threaded over the pegs in ways that it could never be through holes.

How bed-cording patterns evolved in response to the failure of alternative patterns to be economical (of either material or time) is but another example of how failure influences the evolution of artifacts generally, and the forces at work both to drive and to hinder their evolution are most evident in the most common of objects. Thus, the displacement of steel by aluminum as the metal beverage container of choice has been tied to, more than any other factor, the economics of mills, those small fractions of a cent that are saved when each can is made a fraction of a mil thinner in a mill that is producing more than a million a day.

There are countless examples of mass-produced products whose form has been altered to greater or lesser, better or worse degrees because of the demonstrated or perceived economics of changing materials or the processes by which they are altered or assembled. Design and redesign are always comparative activities, with choices necessarily being made to take this over that, these over those, and the choice is generally made in favor of the design that least fails to meet whatever collection of criteria the decision-makers employ. This may be less evident in the case of larger engineering structures or systems, where evolution often takes place on the drawing board and out of public view. In the last century, for example, when the railroads were being extended across the country, there was a continual need for laying out long sections of track over an ever-changing terrain. The route chosen through the wilderness not only determined the gradients up which locomotives would have to pull

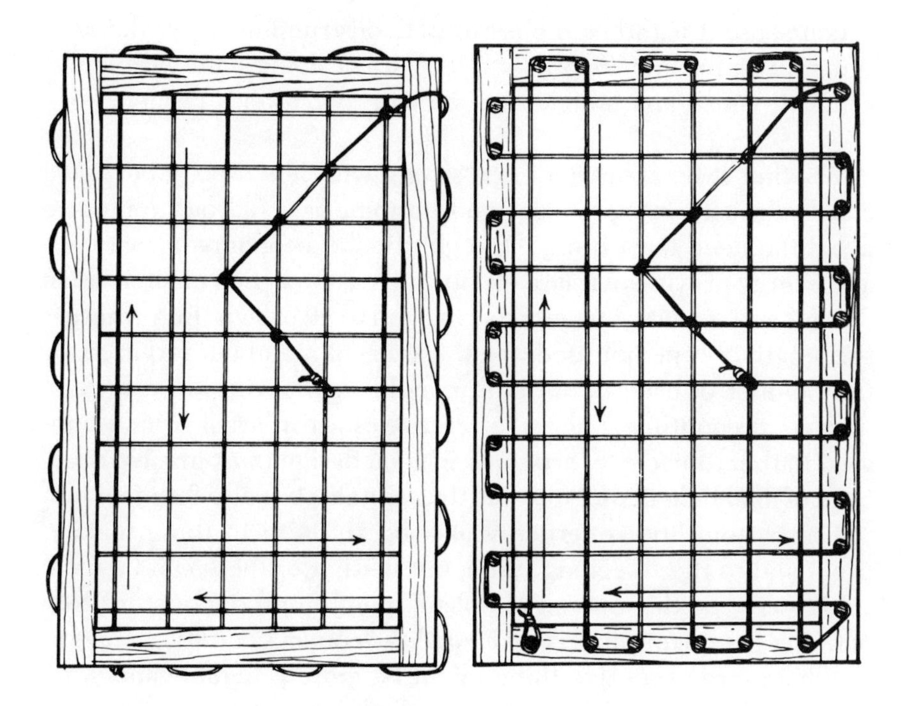

Two schemes were popular for installing rope in Early American beds. In one scheme, the rope was passed through holes in the frame; in another, the rope was passed over pegs. Though there was little difference in the amount of rope used, the two schemes differed in the amount of time required to complete the process. Questions of economy and efficiency have always influenced the nature of engineering designs and the form of artifacts.

but also the number of waterways and valleys that would have to be bridged. This in turn affected the way in which the railroad altered the natural landscape. The characteristic forms of American as opposed to continental railroads—their different gradients, and the use of wood rather than iron in bridges—resulted from the different philosophies of their railroad engineers. The importance of the decision of where to locate a railroad was put in a concise context by A. M. Wellington in his classic *Art of Railway Location:*

> It would be well if engineering were less generally thought of, and even defined, as the art of constructing. In a certain impor-

tant sense it is rather the art of not constructing: or, to define it rudely, but not inaptly, it is the art of doing well with one dollar, which any bungler can do with two after a fashion.

Whether it be an ancient bed strung with less rope, a beverage can made with integral sides and bottom, or a railroad routed to avoid the need for a bridge, the question of economy of materials and energy provides a relatively objective comparison of alternative designs and is central to engineering and to all design. But, whereas the length of rope not used, the thickness of metal not required, or the number of bridges not built may be easy savings to calculate, such bean counting is not what distinguishes an artist from a bungler. Rather, the idea of economy in good design must involve questions of final benefit, not only for the capitalist but also for mankind.

The bottom line is certainly of concern, both to those seeking profit and to those seeking value, but neither of these can be measured solely by the amount of dollars spent on production or product. The nonquantitative word "quality" conveys countless ways in which a more expensive thing might be more profitable and yet a better buy as well. The advantages of thicker metal in an automobile body can clearly be argued from various points of view, including resistance to denting and even simple snob appeal. Whereas the manufacturer can use these as selling points and also as justification for a higher price tag, the buyer can easily justify spending more for a car that will keep its appearance longer and provide a status symbol.

Even when two virtually identical products are available at different prices, it is rare that price will be the sole criterion for choice. Consider the case of food in supermarkets. It is evident that the same product is priced differently in different stores, but not all products are more expensive at store A than at store B. Ideally, someone making shopping decisions based purely on price would compare the prices of each and every item on a shopping list and choose one from store A and another from store B, according to which is the better buy. This is essentially what supermarkets do in reverse (drawing up the shopping list after designing the register tape) when they advertise that a bag of groceries at their store saves the shopper so many dollars and cents over the identical bagful at another store. The advertising claim may certainly be true for the specific selection of items in the bag, but the savings might be reversed with another shopping list. Doing such detailed comparison shopping is obviously time-consuming, and would take the individual shopper at least three trips to two stores to accomplish. The time investment may be worth it to the store manager who uses it as an advertising gimmick over the competition, but how much is the shopper's time worth?

In rare instances, in the days before computerized pricing, different boxes of the same product were sometimes found priced differently on a store's shelf, perhaps because the stockperson neglected to put new higher-price stickers over the ones on the older merchandise, or because the store manager chose not to include subsequent price increases on stock bought at lower wholesale prices. All other things being equal, the consumer would have been foolish not to select a lower-priced box, but all things are seldom equal. Aside from the question of freshness, older packaging may not be as attractive as the newer, and older packaging may not have the same convenient features as newer. Which box the shopper will select depends upon a complex set of criteria that differ from person to person precisely because we all have different priorities with regard to what we consider important (and ultimately economical), and the manufacturer or distributor of consumer products not only can play on this fact but must do so to remain competitive. It is not just capitalism that works on this principle; even in countries where there are long lines for indistinguishable products, the shopper makes a choice whether to stand in this line or that.

The dynamics of food shopping is but a paradigm for making choices among artifacts generally, and although price can often be a major consideration in choosing one box or brand over another, it is seldom the only one. All we need do is browse among the supermarket shelves and read the claims of their "new, improved" stock. Whether one is attempting to sell a new brand of soap to a consumer or an invention to a patent examiner, comparisons with the prior art are central to the case. A new product that is in every way identical to the old except in price is rare, for offering something at a lower price means that somehow it incorporates new materials or ingredients or processes the old in a more efficient way. Speeding up a production line may be a cost saver only if the money saved in production does not have to be spent on advertising to sell the additional output. The demand for some products does become so strong that new plants must be built and advertising hardly seems necessary, but with new plants often (though not always) come new (improved) processes using new materials. Who can remember a wildly successful mass-produced item that maintained all the identical qualities that launched its success? Subsequent versions tend to incorporate features that overcome prior faults or else, when driven by ill-advised economies, introduce new ones.

Inventors may all dream that in the not-too-distant future a new factory will be required to keep up with the demand for their new gadget, but when they are in the early stages of seeking a patent their thoughts are often on past competition. The inventor Nathan Edelson lives in Montana and has worked on designs combining computer work stations with exercise equipment. In the course of describing a trip to the Patent Office in Washington to do his own background checking of what had been done along the lines of his idea for an adjustable or "active" desk, he related his joy at finding that, although there were plenty of precursors to his idea, his invention did have a competitive advantage:

> Ideally in a patent search, you hope to find "prior art" that attempts to accomplish the same "objects" as your invention, but which, for one reason or another, fails to do so adequately. This generally indicates that the potential benefits of your invention are recognized as legitimate, but that the means for achieving them need significant refinement or rethinking.
>
> . . . In the case of my active desk search, I am fortunate. One of the chief "objects" of my invention is to permit users to reposition the desk height quickly and easily so they can avoid postural fixity, which causes musculoskeletal stress. The patents I review indicate that many other inventors have developed adjustable desks, but in every case the movement mechanisms they employ are slow, complicated, and expensive. My desk has an adjustment mechanism that suffers from none of these deficiencies.

Whether or not a patent examiner would share Edelson's judgment that his desk-adjusting mechanism would work faster, operate more simply, and be more economical to manufacture than those in existing patents would have to wait for the patenting process to proceed. However, since Edelson believed that his desk "also offers other novel and useful features," he was optimistic that he would "have a good chance to obtain a valuable new patent."

The potential value of a patent is not often far from the minds of inventors; nor is its cost. In addition to the expense of traveling to the Patent Office or hiring a Washington-based patent agent to search the files for prior art, there are the filing and other costs, which can amount to over $500 for an individual and about twice that for a large corporation. A patent search can be arduous, with or without an adjustable desk, for the incremental improvements in artifacts and processes recorded in the United States alone have given rise to five million patents through 1990. There are movements to computerize the files, but searching must still be done within about seventy thousand classes and subclasses. U.S. patent information is now being made more and more accessible by computer, but in late 1990 it required two compact discs to cover a single week of patents as they were issued. Nine discs were needed to contain the abstracts alone of all the patents issued in Japan during the 1980s. The Patent Office is working to have all U.S. patents available on optical discs, but the process is progressing slowly. Even if computerized, the patent files will be cumbersome to search, for a single subclass could occupy about a thousand compact discs.

Should Nathan Edelson gain a patent for his "active desk," that in itself will not prevent a similar adjustable desk from being made someday. What drives inventors and corporations to carry out laborious patent searches and to bother at all with the details of filing an application and seeing it through the seemingly arcane process is the legal right to sue for infringement. Though some individuals go through the patenting process merely to experience it and to be able to enjoy the achievement of owning a patent, most patents are taken out for their potential economic rather than intellectual value. If, for example, the Edelson Desk does someday become the office desk of choice, there would no doubt be a host of imitators offering look-alikes at lower prices. They would be able to do so because, in addition to saving on research-and-development costs, they might not use as heavy a piece of wood for the top, as thick a piece of Formica for the surface, or as attractive a piece of trim for the edges. The other desks might have a slightly different slant, but they might look close enough to the Edelson Desk (perhaps as seen on TV?) to capture a sizable part of the market.

Edelson, in the meantime, might have just completed a new factory, whose output he has difficulty selling. For all the thought and time he had put into inventing and bringing to market a good solid adjustable desk, perhaps, after some rough starts that required him to seek further capital and to modify his design, he might be left holding an empty cash bag. If he could, however, argue in court that one or more claims in his patent were infringed upon, then he might at least recover something for his efforts. If, on the other hand, in trying to make a less expensive desk than Edelson's, his competitor had come up with a functionally superior and novel design that did something Edelson's desk failed to do, then Edelson would have lost a battle but the world would have won a new form.

The potential value of a patented invention was demonstrated late in 1990, when one inventor won a $10-million settlement against an automobile manufacturer he had approached years earlier with a new idea for windshield wipers. Robert Kearns was a professor at Wayne State University when he reflected upon the failure of existing windshield wipers to work effectively in light rain and drizzle. The wipers had repeatedly to be turned on and off if the driver did not want to be annoyed by the rubbing and streaking as the blades passed over just a few drops of water on each swipe. For some drivers the major annoyance was the distracting noise of the stick-and-slip action, while for others it was the thought of wearing out blades unnecessarily. Some drivers may even have been oblivious to the inefficiency of the wipers in a light rain, and many may simply have accepted the need to flip them on and off as just the way things worked.

Kearns not only noticed the failure of an existing thing to work effectively under all conditions but figured out a way to solve the problem. He invented a mechanism that would allow a variety of settings to sweep the blades across the windshield intermittently, clearing the drops of water only after enough had accumulated for the rubber to work smoothly but before they got too dense to see through safely. The inventor installed his device on his Ford and drove it to the auto manufacturer in Detroit, where engineers seemed immediately to see the advantages of his improvement and asked questions about it. Kearns took their interest as an indication that Ford would buy his invention, and so he expected to be rewarded for his ingenuity.

But when Ford began to install intermittent wipers on its cars without offering Kearns any compensation, he sued for patent infringement. The company's defense was that the idea for such wipers was conceived before a patent was issued to Kearns, and so no infringement had occurred. But, following twelve years of litigation in the courts, Ford agreed to pay the patent holder a settlement, which after legal fees amounted to a royalty of thirty-three cents for every one of the twenty million Fords, Lincolns, and Mercurys that had been manufactured with intermittent windshield wipers. Legal battles with nineteen other automobile manufacturers promised to add to Kearns's ultimate profit from his invention.

Though the more complicated windshield wipers certainly have not made automobiles any less expensive, they have provided such a clear advantage by eliminating some failings of the older, continuously running devices that the overall operation of the automobile is safer and more effective. The visual and aural experience of driving a car through a light rain has certainly been altered, and in a broad sense the car itself and the traffic patterns of which it is a part work more efficiently. And now windshield wipers themselves, as well as their blades, are certainly used more frugally.

13

When Good Is Better Than Best

Just as investors speculate on the future price of oil and other commodities, so do entrepreneurs, venture capitalists, and corporations speculate on the future of new designs. And just as oil prices can depend upon a host of cultural and political factors well beyond the seemingly simple rules of supply and demand, so can the acceptance or rejection of a new or even a modified artifact depend upon much more than how well or poorly its form suits, let alone follows, its function. Indeed, the investor in design is ill served by an adviser who looks too narrowly at technical indicators to prognosticate performance in the marketplace. Case study after case study warns us that no design is sacred and that form follows where the future leads.

As examples like the aluminum can and plastic bottle make so clear, it is not only consumer products proper but also the design of their packaging that can be subject to the times. In the early 1970s, the McDonald's Corporation was encircling its Big Mac in a paper collar, wrapping it in paper and foil, and then inserting all of this in a red box. Such an elaborate package, though hardly an organic form following from any single function, was developed to meet the several functions of getting an elaborate hamburger from behind the counter to the customer's mouth without its looking or feeling like a cold, soggy mess, at least before the first bite. The paper collar kept the double-decker Big Mac from being skewed or squashed in all the wrapping and handling, the paper absorbed excess grease and thus prevented unsightly drips, the foil not only kept the hamburger from becoming cold and dried out but also covered any grease spots on the paper and thus prevented any unsightly appearance from causing Big Mac purchasers to lose their appetites. Finally, the box kept the wrapping from coming undone and gave the Big Mac a special gloss to accompany its special sauce. Even if it was effective, the elaborate packaging took considerable time to assemble and a not inconsiderable amount of time to open. In short, the medium failed to convey the proper message for a fast-food restaurant.

In 1975 McDonald's introduced a new packaging design that seemed to remove all the failings of the old. Each Big Mac was to be packed in a polystyrene "clamshell," an ingenious device made from foamed petroleum products that enabled the hamburger to be packaged in a single motion in a single container that could be opened just as quickly and easily by the consumer. As a bonus feature, customers found that the opened lid of the clamshell provided a convenient bowl for French fries. Moreover, the box evoked the faux-mansard roofs of the McDonald's restaurants and seemed to be the perfect metaphor for the fast-food chain.

The new hamburger packaging was not wholly a new idea, for the same material had been used in the familiar foam egg cartons that were then becoming ubiquitous in supermarkets, but the fast-food application seemed brilliant. The rigid plastic-foam container kept the temperature and moisture in, absorbed grease without becoming unsightly or soggy itself, and provided a neat, colorful, and distinctive one-piece package for the Big Mac. Furthermore, by the mid-1970s there was a growing concern over the profligate use of paper as packaging, and the clamshell thus seemed to constitute an environmentally innovative approach.

The Big Mac clamshell was hailed by designers as a model achievement, and eventually other McDonald's products were being sold in similar packaging, with the clamshell appropriately colored and printed to distinguish, say, a Quarter Pounder from a Quarter Pounder with Cheese. In time, the basic design evolved into a related product that looked somewhat like one flat-opened clamshell covered by another. This provided a divided package that was integral to the marketing of a new sandwich, the McDLT. One compartment of the double polystyrene shell kept a hamburger warm, and the other kept the lettuce and tomato cool, until the customer was ready to combine the ingredients.

When a still newer sandwich, the McChicken, was introduced, it was packaged in a modified clamshell that emphasized one disadvantage of the original design, which seems to have been over-

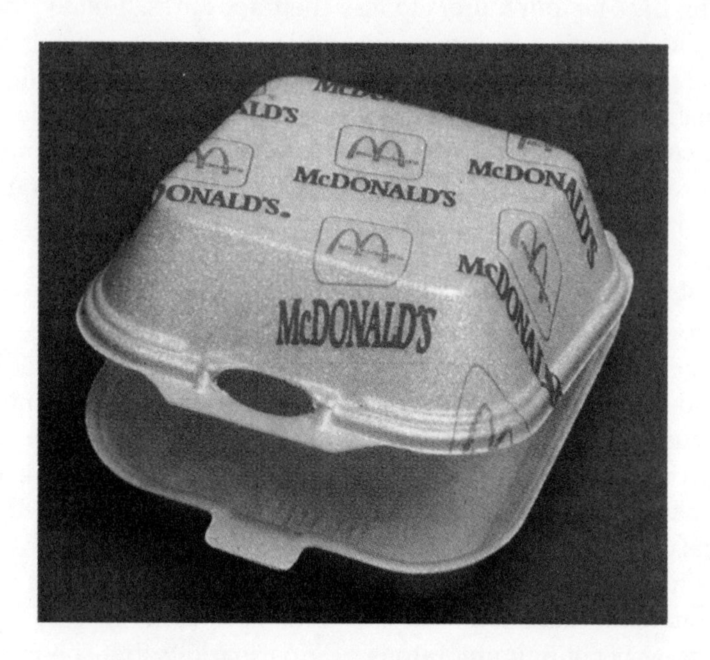

When first introduced, the McDonald's clamshell appeared to be the ideal packaging for a fast-food hamburger. The polystyrene-foam box not only kept the heat and moisture in but also absorbed neatly any errant grease. Furthermore, the hamburger could be boxed in one quick flip of the lid and opened just as easily. Unfortunately, what was once heralded as a brilliant piece of packaging became such an environmental nightmare for the restaurant chain that it reverted to paper packaging.

looked amid all the hoopla accompanying the introduction and acceptance of the plastic box—namely, that it was not very easy to get the Big Mac or Quarter Pounder out of the deep half-shell in which it sat. To make the clamshell much bigger than the hamburger would have made the food look skimpy, yet the fit was so tight it was hard for diners to get their fingers beneath the sandwich, and it was necessary to tip the package over for access to its contents. The new McChicken was packaged in a modified clamshell whose bottom was tapered down from the hinge to the latch so that, when the box was opened, one side of the sandwich was exposed to the fingers for easy removal. This clear improvement in the basic design eliminated the little annoyance of its predecessors but was not adopted for other McDonald's sandwiches, presumably because there was a reluctance to tamper with the "classic" designs that had become so familiar. But, however familiar, these same designs that once seemed so successful from certain functional perspectives, soon came to be seen as failures from other ones.

Within a decade of its introduction, the clamshell began to be attacked as a symbol of wasteful packaging and a threat to the environment. Paper was still a problem, of course, but plastics were perceived to be a worse one. The chlorofluorocarbons (CFCs) that were used in forming the plastic-foam container were implicated in the depletion of the earth's protective ozone layer. McDonald's showed itself to be responsive to environmental concerns by switching to plastic packaging made without CFCs, and the phase-out was completed in 1988. In 1990 the restaurant chain was highlighting the decision in corporate promotional material, stating that the move had been supported by environmental organizations and the Environmental Protection Agency. But even if environmental groups did concur with McDonald's efforts in behalf of the ozone layer, other differences were not necessarily resolved.

The polystyrene clamshell had a useful life of only the short time it held a sandwich from the counter to the table, and the seemingly eternal afterlife of the package made it a very visible contributor to growing litter and pollution problems. The clamshell failed to satisfy environmentalists because it was not biodegradable and it bulked out the contents of landfills. By the late 1980s, the continued criticism of its packaging by environmental activists led McDonald's to explore the possibility of recycling its plastic food containers, but there was skepticism as to whether such an effort was economically feasible. The polystyrene clamshells were often held up as the most visible symbol of profligate disregard for the environment, the commingling of differently processed polystyrene salad bowls and lids, polyethylene-coated paper cups, polypropylene straws, and other fast-food packaging and accessories made it difficult to separate them all for recycling. Furthermore, cleaning up the waste was problematic, compacting it was messy, and storing it unwashed and in bulk was malodorous and space-consuming. Finally, in 1990, the corporation declared that by the end of the year it would begin to phase out the plastic packaging in favor of paper.

The McDonald's plastic clamshells had accounted for about 10 percent of the sales of Amoco Foam Products Company, a division of the oil corporation, and for about 7 or 8 percent of the one billion pounds of foam packaging manufactured in the United States each year. McDonald's was able to make what seemed to be an overnight change in policy because, as it came under increasing attack by environmental groups, the corporation had for some time been weighing the pros and cons of paper versus plastic packaging. In announcing the change, the company's president posed with the director of the Environmental Defense Fund, behind a table crowded with tall piles of foam boxes and the much more modest pile of paper that would replace them. But environmentalists were by no means unanimous in hailing McDonald's decision. Although the Environmental Action Foundation noted that the "polystyrene production process is polluting and the styrene monomer is a suspected human carcinogen," a scientist from the National Audubon Society received the food chain's announcement with less enthusiasm, pointing out that paper was also a pollutant.

Others used the occasion of the packaging-design change to make a further point. In the wake of McDonald's announcement, one of its arch-competitors took out full-page newspaper ads declaring "Burger King applauds McDonald's for its new environmental consciousness." But, the ad continued, "Welcome to the club. We wonder what the planet would be like if you had joined us in 1955?" Nineteen fifty-five was the year since which the then newly created Burger King had used mainly paper packaging. Polystyrene coffee cups were an exception, and in late 1990 they were in the process of being replaced by thick-paper cups.

All of these decisions were clearly more politically than technologically driven, pointing up the complex dynamics behind the evolution of artifacts. The conventional wisdom is that technology affects society in irreversible ways and that, as Ralph Waldo Emerson wrote in a poem, "Things are in the saddle, / and ride mankind." However, we might also extend the metaphor by recognizing that we are capable of rearing up and bucking off things that we find too burdensome or that we feel are taking us in the wrong direction. But, in spite of the spectrum of forces at work in pushing and pulling the form of everything from plastic packaging to the hamburger it contains, there remains a unifying principle behind all influences on form. That principle is embodied in the concept of failure, whether in regard to the technological function of keeping the hamburger fresh and warm or the social function of achieving a healthy and clean environment. The failure of a particular package to perform either of these functions can introduce forces toward change or redesign. But, as the example of hamburger packaging so neatly compresses into the span of a decade and a half, what connotes failure one year may not do so fifteen years hence.

Our collective political memory may understandably seem not to be so long as even four years; for all its supposed objectivity, our technological memory can seem just as short and as subject to slogans over substance, to promises over proof. It was, after all, a rather objective judgment that the foam clamshell did for the Big Mac and McDLT what paper packaging could not. In announcing McDonald's environmentally responsible decision, the company president had to admit that the new containers would not retain heat as well as foam. According to one report, he said that "improvements in cooking methods since the company last used paper-based packaging in the early 1970s would compensate." He also said that "the technology of the cooking process has caught up to the defects of paper," but certainly the truth of that assertion is going to boil down to a question of taste. As for the McDLT, whose very concept relied on the dual-chambered foam package, that was admitted to be "a very difficult problem" indeed. In fact, the McDLT was unavailable while new packaging was being developed.

There are a lot of difficult problems in design, and their solution necessarily depends not only upon where designers understand the problems with the past to be but also upon how clearly designers see the road to the future. The operators of wheeled vehicles, by their very nature, are forward-looking; the earliest carts were pulled rather than pushed by people who could see the path unobstructed before them, perhaps in imitation of the way they pulled plows, and the advantage of this arrangement is clear to anyone who has tried to put a cart before a horse or backed up a car with a trailer attached. In time, people were replaced by draft animals, of course, and the only kind of wagonlike vehicles that seem to have evolved with their prime mover following rather than leading has been human-powered.

For a long time in China, where there was such an effective network of waterways, roads and wheeled vehicles did not evolve into so sophisticated a technology as they did in the West. However, one means of land transportation, the Chinese wheelbarrow, which

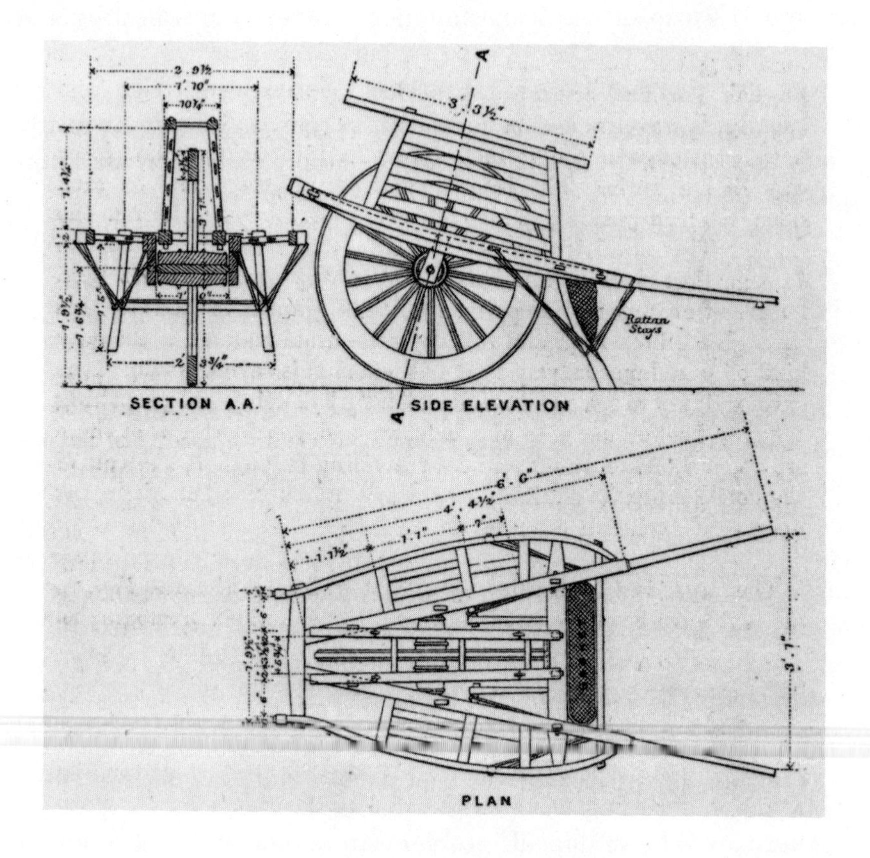

The Chinese wheelbarrow is so constructed that a massive and bulky load may be lashed to the frame in such a way as to achieve almost perfect balance about the large enclosed wheel. Thus loaded and balanced, the wheelbarrow presents little burden to the handles, and the person pushing it can concentrate on maneuvering the vehicle.

is believed to have appeared about eighteen hundred years ago, did develop into a rather ingenious configuration. The Chinese wheelbarrow has a very large wheel, of three-to-four-foot diameter, which is set close to the center of the vehicle. The wheel's upper part is enclosed in a wooden framework onto which an enormous burden can be piled and lashed in a careful arrangement that balances itself both side to side and front to back, so that the pusher is little borne upon by the burden and can concentrate on guiding the barrow.

This vehicle is said to have evolved from two-wheeled carts that were ineffective in rice paddies, whose dry boundaries, on which the wheels worked best, were often little more than narrow-topped embankments. A single-wheeled vehicle could be negotiated along the flattened tops of embankments on which a double-wheeled one could not, but even the single wheel of a pulled wagon could easily slip off the narrow divide unless the puller was exceedingly careful and constantly looked back over his shoulder. Keeping the path in view before the wheel was thus the way to proceed.

Beyond also having a single wheel, the Western wheelbarrow bears little resemblance to its Chinese counterpart, and appears to have developed entirely independently, from a wheelless hod carried stretcherlike by two men. Used in mining and construction work, where narrow passageways and temporary bridges were the rule, the two-man hod was essentially a box with handles extending front and rear. Though this hod was perfectly effective for moving its burden relatively short distances, its great disadvantage was that it could not be operated by a single man. But adding a wheel between one set of handles removed this fault, for now a single man could move the load, bearing no more of the burden than before. The two-man hod was no doubt guided by the leading man, and so the earliest Western wheelbarrows might have been pulled rather than pushed. But the disadvantages of doing this along a narrow plank would have been as evident as those of doing it on the ridge between rice paddies, and so the manner of pushing the wheelbarrow, albeit in an awkwardly stooping fashion, to give the pilot-navigator maximum visibility of the path his wheel was taking, would naturally have evolved.

Looking forward is indeed the essence of design, but artifacts take on their form over the course of long, rough, and frequently precarious roads. When the first horseless carriages were developed, the choices were at least as numerous as those associated with laying out the parts of a motorcycle on a bicycle frame. The designers of the first autos naturally focused on the most innovative aspect of motive power, and did not overwhelm themselves with choices of how to steer the vehicle, whose chassis was still basically a wagon. The role of reins, for example, was played by a lever that extended into the driver's hand.

The Western wheelbarrow appears to have developed from a wheelless hodlike device that was carried by two workers. This illustration from Diderot's *L'Encyclopédie* shows such a hod in use. The singular disadvantage of this kind of hod, that it required two workers simultaneously, was clearly overcome by adding a single wheel between one pair of handles.

With a successor to the horseless carriage firmly established in the automobile—and when roads had been adapted to it rather than it to they—the attention of designers could focus on the details of how it was made and functioned. The American system of manufacture, whereby everything from pins to pistols was either mass-produced by machine or assembled in machinelike fashion, naturally led a Henry Ford to see automobiles manufactured that way. The design of cars was a question of seeing clearly ahead on the road along which cars and the country were heading. All innovators believe they see the road ahead clearly, of course, but on the journey of design all roads fork and fork again into the undergrowth. Which will become the roads more traveled by will depend on style and conformity that designers, no less than poets—if only in retrospect—may lament. And if the choice of which road to take is not obvious, then the shape of the vehicle to travel upon it may be even less so.

The streamlining of airplanes followed naturally from their failure

These two wheelbarrows, illustrated in Agricola's sixteenth-century treatise on mining, clearly have a strong resemblance to the two-man hod depicted two centuries later in *L'Encyclopédie.* While the wheelbarrow clearly had an advantage over the hod in requiring only one person to transport it, the hod retained an advantage over the wheelbarrow when it came to being emptied onto an elevated work space. Such relative advantages and disadvantages among artifacts lead to diversity rather than extinction.

to move efficiently through the air, but the design of the first Wright aeroplanes concentrated rightly not on style but on the principal design problem of the day—that of controlling the craft. With increasing mastery of that came increasing speed, which in turn raised the drag on the boxy shapes whose aesthetics were of little concern in the rush to human flight (a phenomenon to be repeated seventy years hence in the Gossamer Condor). By the 1930s, the teardrop shape, known since the turn of the century to be the form of least resistance, was incorporated into Boeing and Douglas aircraft, and, being the contemporary artifact that best symbolized the future, the airplane set the style for things generally. The most static of mundane objects were streamlined for no functional purpose, and chromed and rounded staplers, pencil sharpeners, and toasters were hailed as the epitome of design.

Streamlining American automobiles began with some subtle changes introduced in the 1920s, but the solidly established squarish Fords set the aesthetic standard. Radical streamlining, such as introduced by Buckminster Fuller in his Dymaxion car exhibited in 1935 at the Chicago World's Fair, was clearly "futuristic," and hence not taken as seriously as cars of the present. The sensibly streamlined 1934 Chrysler Airflow rounded and tapered the boxy profile, fenders, and windows of contemporary designs, but it was not a commercial success. The immediate postwar period, which the atomic bomb, if nothing else, defined as the future realized, saw the arrival of truly streamlined cars in the 1947 Studebaker. Though the design owed its aesthetic appearance to Raymond Loewy, he clearly acknowledged the indispensable entrepreneurial role of Studebaker's president in turning sketches to reality. With the arrival of the future, as embodied in the jet-and-atomic age, automobile styling no longer had to hark back to its roots, and the fins of rockets began to ornament the tails of Cadillacs in 1948. Throughout the 1950s, fins grew to amazing proportions, each year's models outdoing the last for no functional purpose other than that the new style sold cars.

With the orbiting of the artificial satellite Sputnik in 1957, the space race had begun, and a new design aesthetic was in place. Fins were necessary on the rockets that launched satellites, but the artificial moons themselves needed no streamlining or stabilizers to orbit in the virtually frictionless void above the earth's atmosphere. Sputnik was a surprise, of course, and so automobile designers could not use it to define their immediately upcoming models; with time, however, the look of the future was toward the moon and outer space. The lunar excursion module was a contraption worthy of the Wright Brothers, and streamlining was a distinct disadvantage to a space capsule returning through earth's atmosphere. Designs for interplanetary probes re-emphasized the curiously boxy beauty of the future, and the space shuttle became the vehicle of design as well as of transportation choice. The silhouette of terrestrial vans introduced in the 1980s bears a distinct resemblance to the nose of the shuttle, and names like Ford's Aerostar leave little to the imagination as to what images they wish to evoke. Automobiles are marketed like hamburgers, and how well the future dreams and detestations of the customer are read, whether in the product per se or in its packaging, can make the difference between commercial success and failure when design must satisfy so many functions that a single form could hardly be expected to follow from them.

Though all design is necessarily forward-looking, all design or design changes are not necessarily motivated by fickle style trends, whether they be in the environmental politics of plastic packaging or in the patriotic images of advanced technology. The best in design always prefers substance over style, and the lasting concept over the ephemeral gimmick. Design problems arise out of the failure of some existing thing, system, or process to function as well as might be hoped, and they arise also out of anticipated situations wherein failure is envisioned.

Ralph Caplan's book *By Design* is distinguished by the intriguing situation described in its subtitle: *Why There Are No Locks on the Bathroom Doors in the Hotel Louis XIV.* Caplan writes of the bathroom-door object lesson as "an ingenious example of the product-situation cycle" and as "the perfect fusion of product and circumstance, and a demonstration of the design process at its best." His language is more that of the industrial designer than of the engineer, but the hotel problem that Caplan highlights is indeed a wonderful model for how designers must always look ahead, to the future situations and circumstances in which their product will be employed—and to how it might fail.

Before it was destroyed by fire, L'Hôtel Louis XIV, which was located on the waterfront in Quebec, advertised private baths. However, their privacy was of a limited and precarious kind, for each bath was located between a pair of guest rooms, both of which opened into it. This arrangement is not uncommon in private homes, where bedrooms share a bathroom or where a bathroom opening into a bedroom also opens into a hallway. In all such situations, the basic design objective is to provide privacy for whoever might be using the bathroom. This can be achieved in many ways, of course, and the most obvious and common way is to have locks on each of the doors, so that the bathroom user may bar others from entering. The failure of this solution is frequent and frustrating: the person who has finished with the bathroom forgets to unlock the second door, causing at least a little inconvenience for the next user who tries to enter it. In bathrooms shared by siblings, screaming through the locked door may or may not get results, but generally there is little more than the temporary inconvenience of having to go around to the other door or to another bathroom in the house. Families that find bathroom doors too frequently locked can remove all locks from the doors and trust everyone to knock before entering.

In the case of bathrooms shared by unrelated guests, the problem is less easily solved. I once stayed in a wonderful old home across the street from Washington University in St. Louis in which two guest rooms shared the same bathroom. Individual guests were expected to come and go at odd hours, and they often wished to leave irreplaceable things like slides and manuscripts in their rooms. Hence, it was desirable that the rooms could be locked against entry from both the hall and the bathroom and, at the same time, that both bathroom doors could be locked from the inside so that privacy could be assured. The arrangement no doubt resulted in many frustrated guests finding themselves locked out of the bathroom, the other guest not in, and the housekeeper nowhere to be found. The measures taken to avoid this situation consisted of a nicely printed sign placed prominently on the dresser beside the bathroom door, reminding each guest to unlock the other guest's door before leaving the bathroom. I am sure I was not the only guest who suffered from the inadequacy of that solution.

Whether it was the repeated failure of guests to remember to unlock their neighbors' doors or some uncommon foresight on the part of the Louis XIV to anticipate the failure of locks to provide fail-safe access to an empty private bathroom, the hotel solved the problem in an ingenious way. Each bathroom door did have a lock on the guest-room side, of course, for otherwise a stranger could come in through the common bathroom, but there were no locks at all on the insides of the bathroom doors. To gain privacy, a guest hooked together in the middle of the room the ends of the three-and-a-half-foot lengths of leather thong attached to each doorknob. Even if the leathers stretched tautly across the bathroom interfered a bit with movement inside, they effectively prevented either door from being pulled open while the bathroom was occupied. However, to open either one of the doors to leave the bathroom, the thongs had to be unhooked, thus unlocking both doors simultaneously.

Focusing too closely on the immediate design problem, whether it be locking bathroom doors for privacy or canning food for preser-

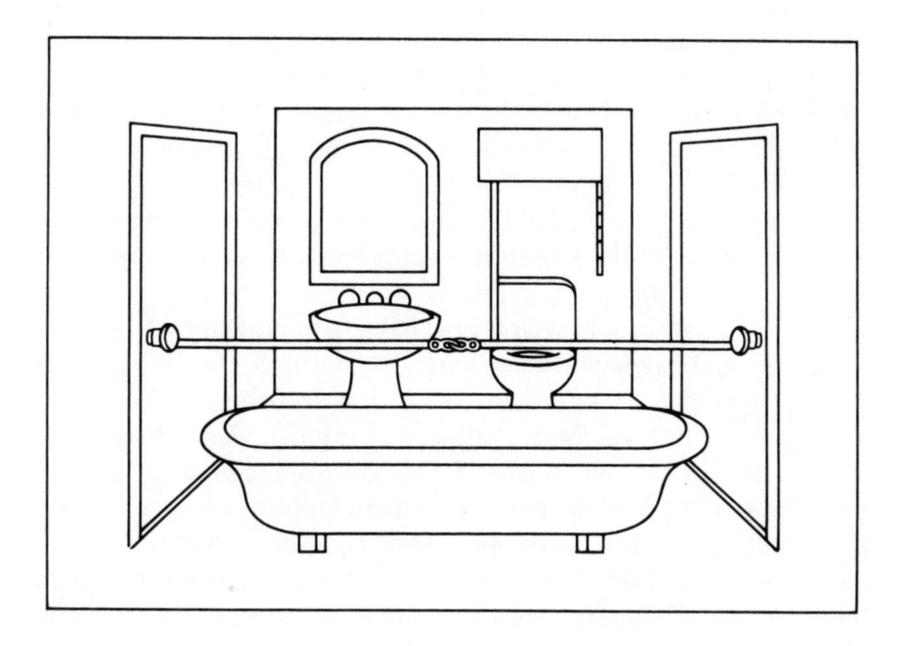

This sketch of one of the bathrooms in L'Hôtel Louis XIV shows leather thongs attached to the doorknobs and fastened together by the occupant (not shown) to ensure privacy. The occupant cannot leave without uncoupling the thongs, and so cannot forget to unlock the door to the room of the person sharing the bathroom.

vation, frequently results in solutions that themselves give rise to more difficult design problems in the future. In the days before plastic was ubiquitous, waste baskets and trash cans were commonly made of metal, and they were emptied by being turned upside down into large collection barrels or bins. Throwing an apple core or a banana peel into a waste basket could leave on its bottom a reminder of lunch that lingered in the office air for days. Disposing of an "empty" can of soda in the waste basket often dripped a sticky mess onto the bottom. In time, a waste basket could become rather crusty and sticky, and washing the metal container, whose finish had become scratched, dented, and worn off during years of being banged about in the emptying process, only caused it to rust and become unsightly. When plastic bags became almost universally used as waste-basket and trash-can liners, they seemed to promise not only a relief from unsightly and unsanitary conditions but also a more efficient and pleasant means for janitors and cleaning crews to empty out the trash. Full bags could just be lifted out of the waste baskets and replaced with clean bags. Larger trash containers in public places were to work in a similar fashion, and there was to be a gain in convenience on the part of both trash disposers and trash collectors. The former would be surrounded by cleaner waste baskets and cans, and the jobs of the latter could be done more easily and conveniently.

In practice, what seems largely to have happened is that the bags have altered the behavior of everyone concerned with their use and disposal, resulting, in some cases, in what is arguably an unforeseen decrease in the level of sanitation and appearance. Because the plastic bags, at least when free of rips and tears, do not leak, many people seem to have become much less thoughtful about what they throw into the trash. Half-empty yogurt containers, half-full cans of soda, and other lunchtime leftovers that might once have been taken to the restroom to be washed down the drain, seem more and more to be tossed away without a second thought. After all, the plastic bag will contain them and be removed before the mold or flies arrive. Many of those who empty waste baskets seem to have developed a different kind of expediency, emptying the waste basket of its contents the old way, by turning it upside down. The plastic bag is not always replaced, perhaps to save on supplies, or to save the time of having to fit another, often ill-fitting, bag over the waste basket, and thus to have more time to spend on other chores or pleasures. As a result, residue can now collect in the bottom of plastic liners, at least those that have not become punctured, and offices may be no more sanitary or fragrant than before.

The situation with public trash cans seems no better. The proliferation of fast and prepackaged food has increased the amount of food-contaminated waste. Since so much of this food and drink is not all that palatable to begin with, plastic-lined trash cans are frequently full of rather ripe and wet garbage. Where squirrels are legion, they often forage about in the trash cans, surprising quite a few passersby with the noises they create inside the dark receptacles and startling more by clambering out as footsteps approach. The squirrels, if not the sharp edges of waste, frequently puncture the plastic bags, which are usually full to overflowing, especially after long weekends. Being so full and foul, the bags are replaced with clean ones, but an early-morning walk in the wake of the garbage pickup reveals a host of sticky trails converging from trash containers to wherever the truck was parked. Because so much of the trash is light and bulky, the garbage truck is equipped with a compactor that allows a good number of bags to be fitted into each truckload, but compacting a plastic bag is like squeezing a grapefruit half, and the liquid naturally squirts about and follows the law of gravity. The truck seems incapable of containing all the liquid, and so it leaks out the bottom and onto the pavement. This having been noticed by the trash collectors, they have come to park their truck over storm drains, so that the bulk of the liquid drops into the sewer. But in dry weather, the slop just sits there and cooks into a foul soup. By afternoon on some days, the stench can be unbearable.

Plastic trash bags that were ostensibly designed to improve the quality of our lives have thus altered our behavior and environment. The malodorous and unsanitary conditions of their disposal aside, the bags themselves seem to be a blight on the overwhelming majority of places, private and public alike. In order to hold their shape and contents, they are folded over the sides of the waste basket or trash can, and it does not seem possible to do this in an aesthetically pleasing way. The bags are often much larger than the receptacle, so that they may be gathered and closed for disposal, but the extra plastic must then be bunched up or extended halfway down the side of the container, curiously reminiscent of the way some old women used to roll their stockings halfway down their legs. Whether the bag is bunched or rolled, however, a waste basket designed to harmonize with a neat and businesslike office or courtroom decor, or a trash can designed to be as unobtrusive as possible in a garden or a tree-lined path, ends up looking like nothing more attractive than some packaging only half removed. By force of quickly developed habit, it seems that virtually every trash receptacle is now lined with a plastic bag, whether necessary or not. In a library that I frequent where food and drink are strictly forbidden, all the waste baskets are overflowing with filmy plastic bags into which nothing but paper is ever likely to be dropped. If ever there was a pervasive design success turned failure, it is the plastic bag, now poised for an evolutionary improvement.

The design of everything from fast-food packaging to litter containers must look beyond immediate use. Each artifact introduced into the universe of people and things alters the behavior of both.

Whether the alteration be malevolent or benign is not always evident at the outset, but its impact certainly can be better anticipated if designers look down the design road and well beyond their immediate objectives. Though the best designs deal successfully with the future, that does not mean they are futuristic. All too often, the uncritical adoption of new materials or devices to solve old or imagined problems can create newer and more complex problems in an altered environment. The look of the future has so frequently become the blight of the present that it behooves designers to look more carefully and thoughtfully beyond appearances and short-term goals to the substance of designs and their long-term consequences. The analogy with business is that one must look beyond the quarterly bottom line and think in terms of the company history that will one day be written.

14

Always Room for Improvement

In a column entitled "March of the Engineers," the humorist and social critic Russell Baker lamented the complexity and sophistication of his office's new telephone system. Not only did everyone have to attend classes for instruction in how to use it, but such features as call forwarding seemed to Baker to be taking technology too far: he wanted to be able to travel to distant places and not have his telephone calls follow him around the world. Baker closed his column by defining the new telephone system as "another bleak example of the horrors created when engineers refuse to leave well enough alone."

Every technological change has the potential for being both cursed and praised. What seems "well enough" to one critic may seem vastly deficient to another, and the roles of the critics may reverse from time to time, from situation to situation, even in the same individual. In the case of call forwarding, for example, another reporter might find it a wonderful feature were he trying to track down someone to confirm a detail in a story whose deadline was fast approaching.

Russell Baker is not the only observer of late-twentieth-century technology who has lamented a new telephone system. In *The Design of Everyday Things,* Donald Norman wrote that "new telephone systems have proven to be another excellent example of incomprehensible design." Indeed, elaborate push-button telephone systems provide a virtual paradigm for Norman's inquiry into modern devices that "add to the stresses of life rather than reduce them." He could "count upon finding a particularly bad example" of a system wherever he traveled, and many of the anecdotes he relates ring true to anyone who has gone through the trauma of adapting to a new instrument on the desk.

Our school recently got its own sophisticated new telephone system, and many of my first reactions were similar to Baker's and Norman's. I resented losing my familiar old black rotary-dial instrument, with its single row of extension and intercom buttons whose code I had grown to understand. In time, however, I also remembered the frustrations when I first had to deal with that piece of nostalgia, and then I considered some of its failings that the new system corrected. The old black phone had been connected with dozens of like phones through but three outside lines, and only one of them had long-distance capability. When I wanted to make a phone call, I frequently had to wait for one of the lighted buttons to go off and hope I could lift the receiver to get a dial tone before one of my colleagues did. If I were to misdial what then seemed to be endless digits, or if I got a busy signal, I risked losing my line to another. Since the new telephones have been installed, I have never had to wait for a line, and I have learned the convenience of such features as automatic redialing, in which I need only push a single button to have a long string of digits repeated, or automatic callback, in which I need only push another button to have my phone ring when the line that was busy is free.

As for call forwarding, my phone has that also, but I have yet to use it to forward my calls to the beach in August. Rather, I have employed it to send calls to our department secretary so that she may take messages or handle business when I cannot or do not wish to answer my phone. My new phone also has a voice-mail feature, which at the press of a button stops my phone from ringing and at the same time activates an answering system that records messages I can listen to and respond to at my convenience. Baker's new phone may have even more features, and he is free to use or ignore as many as he wishes. From my point of view, the engineers have made well enough better and given me the choice of taking it or leaving it.

I will admit that the new phone was somewhat intimidating at first. Its buttons were unfamiliar, and the options seemed overwhelming. I also resented having to stand with a crowd of my colleagues around a telephone representative going too quickly over features and using jargon she was too familiar with and we were generally too proud to ask about. I suspect that not a few of my colleagues eventually learned to work their telephones feature by feature, as did I, by spending hours in the privacy of their offices poring over the always confusing and often contradictory instruction manual. When any of us mastered some esoteric new feature, he would allude to it in lunchtime conversation, finding it gratifying when he turned out to be the first to make use of that item; correspondingly, each of us feared the embarrassment of learning that he was the only one still stymied by some other arcane detail.

Ambivalent feelings toward evolving technology are nothing new. I recall that when push-button telephones were first introduced I scoffed at them. Thinking, naïvely, that the single purpose of push buttons was to enable one to complete phone calls faster, I ridiculed anyone who did not have the time to turn the rotary mechanical dial through seven digits to call home. But those were my salad days, when time seemed to move more slowly and telephone numbers were much shorter. I was then still in awe of the simple fact that I could dial a bunch of numbers and cause a telephone in another state to ring. My finger became accustomed to the unnatural but not unpleasant motion of dialing, at least before arthritis cramped my style, and I wondered who would need to dial a telephone in any other way or any more quickly. But now, having touched a push-button telephone, I find it difficult and sometimes downright annoying to have to turn the rotary dial on some of our telephones at home. It seems to take an eternity for the dial to return to my waiting finger after I have cranked a "9" through more than 270 degrees.

Why do what prove in retrospect to be such obvious technological advantages put some of us off at first? It seems in part to be a matter of familiarity's breeding content, at least when it comes to inanimate artifacts whose form our hands have often grown to glove. The appearance of a new form, possibly accompanied by new functions, is intrusive and threatening. After all, a technological artifact like the old black rotary-dial telephone had been raised to the status of a cultural icon. Without thinking, we could use it and watch it being used. It had long ceased to be conspicuous, but let a movie actor dial a phone number with only six digits, his finger in the same hole for all, and the verisimilitude of a whole scene could be jeopardized—unless, of course, his blunder had meaning in the plot. The introduction of the push-button phone seemed to end all of that, and it took some of us quite a while to acknowledge that it had given us something in return. The electronic tones that accompany pushed buttons have become as familiar as the ratcheting advance and clicking return of the old rotary dial, and sometimes they sound like snatches of favorite songs. I have come to find some pleasure in being able to push the buttons in a staccato fashion, and the faster the more satisfying. Telephone numbers have come to take on a visual quality, and I can remember some only by the distinct pattern my finger hopscotches out on the keypad. My automated teller machine access code has a predominately horizontal pattern, and my voice-mail retrieval code has a vertical one; without these visual and physical mnemonics I would have a hard time getting cash or phone messages out of the machines.

The newest telephone systems do not work perfectly, of course, but what does? The evolution of artifacts and their enabling infrastructures—hardware and software, in computer talk—does generally proceed along a route whose milestones read "good," "better," "best," but this last so often appears really to be just over the next hill, as elusive as Shangri-La. The way itself is seldom without its detours, layovers, wrong turns, retracings, and accidents. Especially when the technology is complicated and its goals are ambitious, the road to totally satisfactory performance and acceptance is frequently littered with doubt and second-guessing, with wrecks and breakdowns. At first, neither the designers nor the users of a new technology may fully understand it, and so its progress is impeded and it can cause terrible traffic jams.

Some of the frustrations that Baker has articulated for the telephone have been echoed recently for a host of electronic devices. An editorial in the trade magazine *Design News* discussed some of the editor's irritations with consumer products that it seemed reasonable to expect to be better designed. The editorial "struck a chord" for many of the magazine's readers, who are virtually all designers or engineers themselves, and they responded with their own lists of "aggravating products." Packaging was mentioned by many respondents, who found it to be "too efficient" and "impenetrable." This is a problem as old as nature, of course, as exemplified in the predator tearing at the captured prey, or in the native islander wrestling with a fallen coconut. We have seen that the tin can existed long before an effective can opener, and getting at the product behind much plastic packaging today can be a frustrating and time-consuming experience for an extraordinary number of otherwise adept adults, as is demonstrated on airline flight after airline flight of passengers trying to open their complimentary bags of peanuts. There is really no excuse for designers to make packages so secure that they cause consumers even to remark about them.

Controls on electronic equipment are also a kind of packaging, for unless we can master them we cannot use the product inside the black box. Among *Design News* readers, "the myriad of setting techniques for digital clocks, watches and VCRs" was the "most universal complaint." This is certainly understandable; who has not fumbled by trial and error and jumped through hoops of wires and cords to get some new electronic device to perform its tricks? My own experience has been that, when I master a few moves to get the new clock to keep time or the VCR to record and play, I do very little further exploring of the controls. Thus, I effectively never fully open the package that contains additional features.

In spite of our frustrations with and incomplete mastery of electronic equipment, we have bought it in droves. By 1990 three-quarters of all American homes had microwave ovens and more than 60 percent had video-cassette recorders. Those who do not own such things are the object, if not of ridicule, at least of advertising campaigns, and in these even the electronics companies can acknowledge the problems of their imperfectly evolved products. One company, Goldstar Electronics, on launching a campaign that stressed the "user friendlier" nature of its products, admitted that "the perception among most consumers is that the sophisticated electronic products on the market are difficult, if not impossible, to use," and they wanted to convey the impression that theirs were "designed with real people in mind." In an ironic development for an industry that seems to come up with increasingly complex products, Goldstar wished to differentiate its products from those of its better-known competitors by touting them as "less sophisticated gadgets" that were easier to use.

The basic function of consumer electronic devices, including all of their special features, has seldom been in doubt. A digital watch is intended to tell time and date, to sound alarms, and so forth. A VCR is to record programs and play video tapes, and to provide us with the ability to record one television show while watching another, or to record a show while dining out. Such objectives were clearly incorporated into the design problems out of which evolved the artifactual solutions now on catalogue pages and store shelves. The variety on display there, especially in the configuration of dials and controls, is but further evidence negating the notion that form follows function. Indeed, as we have seen repeatedly, it is precisely the failure of these things to perform as perfectly as someone can imagine that will cause them to evolve through their failures toward "perfection." That is an ironically relative objective, of course, because in the meantime we users are adapting to the imperfections of the existing devices. A thing can never be separated from its users, even in its evolution.

Why designers do not get things right the first time may be more understandable than excusable. Whether electronics designers pay less attention to how their devices will be operated, or whether their familiarity with the electronic guts of their own little monsters hardens them against these monsters' facial expressions, there is a consensus among consumers and reflective critics like Donald Norman, who has characterized "usable design" as the "next competitive frontier," that things seldom live up to their promise. Norman states flatly, "Warning labels and large instruction manuals are signs of failures, attempts to patch up problems that should have been avoided by proper design in the first place." He is correct, of course, but how is it that designers have, almost to a person, been so myopic?

Given the problem of designing anything, from paper clip to microwave oven to suspension bridge, the first objective clearly has to be to get the thing to perform its primary function, whether that be to hold papers together, to cook food, or to span a river. Naturally, designers will concentrate on these things first, and in the process of doing so will become familiar with their designs in ways that few other individuals will ever need to or probably want to be. The original designers of paper clips, for example, will know the ways of the wire they bend first in their minds and then on paper and then in machines. They will learn how some wire cracks when bent into too tight a curve, and how other wire does not lose its spring enough to be formed. In time they will bend the right wire in the right configuration to meet their self-imposed if often ill-defined goals, but more likely they will end up with a bunch of wires in a bunch of configurations, as demonstrated in patent after patent, and as allowed for in claim after claim, each manifestation of which has advantages and disadvantages relative to the others. Out of these, they and their entrepreneurial, manufacturing, and marketing partners in design will select something to make and sell. Though the objective of how the end product will be used is never far from consciousness, those who are involved throughout the design process necessarily become so familiar and friendly with the object of their conception that they can operate it with an ease and care that the uninitiated may never know. An act as seemingly simple as attaching a new-style paper clip to a pile of papers will always be easier for the clip's designer than for the first-time user.

Even if a special effort is made to give a new product over to a human-factors engineer, whose task it is to suggest modifications to make the product user-friendly, the result will only be as successful as the process is complete in anticipating how the product will fail to function. If the engineers tacitly assume that all users will be right-handed, for example, the product may have no chance of being user-friendly for 10 percent of the population. Success depends wholly on the anticipation and obviation of failure, and it is virtually impossible to anticipate all the uses and abuses to which a product will be subjected until it is in fact used and abused not in the laboratory but in real life. Hence, new products are seldom even near perfect, but we buy them and adapt to their form because they do fulfill, however imperfectly, a function that we find useful.

Whether acceptance or rejection is the fate of some new artifact or technological system, the evolutionary process is universally one of relatives and comparatives. Whereas Russell Baker may have cursed the engineers for not leaving well enough alone, what constitutes well enough depends—as it always has. From a certain point of view, prehistoric life was all well and good enough for prehistoric man and woman. Indeed, the artifacts and technology then in existence played a large part in defining the nature of the era. By definition, prehistoric tools and ways were (perfectly?) adequate for getting along in the prehistoric world. The argument that technological advances were necessary to advance civilization is at best a tautology and at worst akin to the myth that necessity is the mother of invention.

What ultimately mandates the fact of technological evolution may be as fundamentally ineffable as what mandates the fact of natural evolution. That is not to say that there is not some dynamic at work but, rather, to suggest that a kind of evolutionary process is inextricably involved with the processes of life and living. Technology and its ancillary artifacts are concomitants to human existence, and it behooves us to understand their nature as well as our own, flawed and imperfect as they necessarily may be. That understanding is most accessible at the microcosmic and microtemporal level, where one thing follows from another as a child from its parent, and the understanding is most acute when it resolves the dilemma of the famous and the obscure, the great and the small, the accepted and the rejected, by explaining their genesis equally while at the same time explaining their divergence of achievement within a common context.

The various manifestations of failure, as have been articulated in case studies throughout this book, provide the conceptual underpinning for understanding the evolving form of artifacts and the fabric of technology into which they are inextricably woven. It is clearly the perception of failure in existing technology that drives inventors, designers, and engineers to modify what others may find perfectly adequate, or at least usable. What constitutes failure and what improvement is not totally objective, for in the final analysis a considerable list of criteria, ranging from the functional to the aesthetic, from the economic to the moral, can come into play. Nevertheless, each criterion must be judged in a context of failure, which, though perhaps much easier than success to quantify, will always retain an aspect of subjectivity. The spectrum of subjectivity may appear to narrow to a band of objectivity within the confines of disciplinary discussion, but when a diversity of individuals and groups comes together to discuss criteria of success and failure, consensus can be an elusive state.

Naturally, the simpler the artifact and the fewer criteria applied to judge it, the less unsettled and controversial may be its form. The paper clip, for example, so unthreatening and controllable, seems easily to attract the admiration rather than the ire of critics and columnists, and appears to be embraced by almost everyone as a little marvel. Who but inventors has thought otherwise? And yet to look closely at this technologically lowbrow artifact is to discover the essence of how even the most elaborate of things evolves. A complex system like a nuclear-power plant, on the other hand, which provides a surfeit of detail at every level and is judged by numerous criteria, including some rather final ones, is a most poor primer on technology. But who should not care about it? Something like a new telephone system is in the middle ground of complexity and consequentiality. Regardless of their level of technology, if the same evolutionary principles govern these artifacts and those in between, then understanding more about any one of them enables us better to understand (and control) them all.

Is all technology for the better, at least in social intention? The simple answer appears to be no, for there seem always to have been among us those who would exploit technology as they would exploit people. Indeed, just as magicians have long employed gimmicks and gadgets to deceive their audiences, so unscrupulous merchants and worse have not infrequently abused technology or played on the trust of their victims in the objectivity of technology. The butcher with his thumb on the meat scale is perhaps among the crudest manifestations of such deception; more sophisticated versions of the same idea have existed since ancient times. Almost twenty-five centuries ago, the Peripatetic author of the *Mechanica* asked why larger balances were more accurate than smaller ones. After answering his own question with an elaborate geometrical explanation involving the properties of circular motion, he explained that dishonest dye merchants preferred small to larger balances because deception could better be practiced: "This is how sellers of purple arrange their weighing machines to deceive, by putting the cord out of true center, and pouring lead into one arm of the balance, or by employing [heavier] wood for the side to which they want it to incline." A slight imbalance in favor of the merchant was magnified by a longer balance arm, and so a smaller device was preferred to escape detection.

But such aberrations in the human use of technology are no more an indictment of technology than criminals should be of the whole human race. Not that designers and engineers, perhaps sometimes in the service of merchants of dyes and worse, do not make mistakes or commit errors in judgment; they do—just as we are all fallible in everything we do. We all make wrong turns confidently, and when this happens the best course of action is to recognize our mistake as soon as possible, pull over to the side of the road, and consult a map to set us right. However, we all know how much easier it can be, especially in the company of others, to continue in the wrong direction than to admit our error and get on with correcting it. Designers and engineers, who after all are people first, can be subject to the same fallibilities, especially when they also suffer from a technological myopia that makes it difficult, if not impossible, for them to focus on several levels of a design problem. A technologically savvy and understanding public is the best check on errant design.

The adaptability of humans to the imperfections of artifacts is perhaps the final determinant in establishing the ultimate form of so many of the things we use, even if with a cursed affection. For all of Russell Baker's griping about a new telephone system, he no doubt eventually adapted to it and perhaps even came to appreciate (without writing about) at least some of the features he once thought so awkward and inscrutable. It is not so much that technology marches inexorably forward and that we risk being left behind if we do not fall into step. Rather, the evolution of the overwhelming majority of artifacts, in both form and function, is fundamentally well intentioned and for the better.

The very fact that we are so adaptable to our artifactual and technological environment is often what makes us resistant to changes in it, especially as we grow older and accumulate our own familiar things and ways with them. Since old telephones did not have features such as call forwarding or voice mail, for example, we could either accept that we would miss calls or take steps to not miss them. A reporter or someone else who depended heavily upon the telephone could be sure that the phone would be answered in his or her absence—by a colleague, a secretary, an assistant, or even an answering service or machine. We did not *need* anything different, but when newer things do become available, some of us can immediately see their benefits. The automatic features on newer telephones have enabled even the free-lance person who works alone at home to have in a single phone all the telephone conveniences of an office worker with a support staff and a network of phones. However, it is the generation that is young enough not to have become so familiar with the old, and yet not so young as to be without the financial resources to do so, that usually embraces the newest technology first.

Whether our sensibilities are with the aging observers of the world or with the up-and-coming generation, the forms of the artifacts that will have an impact on and shape all of our lives are shaped by someone's perception of failure in existing artifacts. That someone will most likely be an engineer, a designer, or an inventor who looks at things in the peculiar way of the technological critic. If the critic has the means to produce a prototype of an improved artifact, or if the critic has the talent of communication or the power of persuasion to involve a corporate sponsor or an entrepreneur to produce it, then the rest of us may be presented with a choice between old and new. In some cases the choice is usurped from us, for manufacturers can have their own criteria of what constitutes failure and improvement, and these criteria involve profit and loss. Thus, what might appear to consumers to be a needed improvement, might appear to manufacturers to be unprofitable. Decisions to make things lighter, thinner, and cheaper may be based no less on perceptions of failure than the decision to adjust a clock that fails to keep time.

The evolution of form begins with the perception of failure, but it is propagated through the language of comparatives. "Lighter," "thinner," "cheaper" are comparative assertions of improvement, and the possibility of attaching such claims to a new product directly influences the evolution of its form. Competition is by its very nature a struggle for superiority, and thus superlative claims of "lightest," "thinnest," "cheapest" often become the ultimate goals. But, as with all design problems, when there is more than a single goal, the goals more often than not are incompatible. Thus, the lightest and thinnest crystal can be expected also to be the most expensive. But limits on the form of artifacts are also defined by failure, for too light and too thin a piece of crystal might hardly be usable.

I once saw a fine Orrefors water goblet broken when a dinner guest offered her small child a drink from it. The child, perhaps used to teething on jelly jars or heavy plastic tumblers, had no respect for the delicacy of the goblet and shattered the crystal into a shower of bite-sized pieces. The suddenness of the accident apparently so startled the child that the broken glass just fell from his gaping jaw. Neither his mouth nor his sensibilities were hurt, but his mother was mortified and my wife and I were left with an odd set of crystal.

The child's mother offered to replace the broken goblet, of course, and a new one was ordered. When the piece arrived, my wife noticed immediately that the new goblet was heavier than the one that had broken, and all subsequent replacements were just as expensive but not nearly so light and thin as those in the original wedding gift. This gift had come at the time when the Orrefors was made as thin as it would get; orders for replacements came in with complaints of excessive fragility. Certainly goblets even lighter and thinner might have been conceived, but then adults too might have had to drink from them with great care, and washing them might have been a rather anxiety-ridden task. The crystal was so light and thin that setting down a wineglass just a little off the vertical on an uncushioned table was sufficient to cause a stem to snap. To make the crystal thinner might have allowed light to play even more delicately upon a glass and its contents, but the usability of the stemware might have been so marginal that it would have been more often than not left in the china cabinet while more hearty water- and wineglasses allowed diners to enjoy their food and drink without risking the crystal or their nerves.

If the world of design is understood to include not only things we can hold in our hands and operate but also the organizations and systems that produce and distribute those things, then we can explain virtually every generation and alteration of any artifact or technological system as being in response to the real or perceived failure of its antecedents to function as expected. But since even real failures, let alone perceived ones, are really matters of definition and degree, what constitutes a useful improvement to one person may represent a deterioration to another. There are countless patents for things that have been labeled new and useful by few others than an inventor and a patent examiner. These things have existed in unique examples in only a few minds, drawings, and perhaps prototypes, but they have been no less reactions to failure than the most successful consumer products.

Jacob Rabinow has related the story of designing a pick-proof lock, certainly an invention that corrected a shortcoming of existing locks. His idea for a more secure lock revolved about a key that was formed from an extremely thin strip of sheet metal, bent into a shape that displaced the lock's tumblers to just the right positions. Typical lock-picking devices, such as bobby pins, would not work, because their very thickness would displace the tumblers beyond the unlocking position. Rabinow was granted two patents for his lock and key, but could not sell the idea to any manufacturer, because the key looked "peculiar." He echoed Raymond Loewy's dictum about "most advanced yet acceptable" designs by attributing to manufacturers the motto "Make it better, but don't change anything."

The inertia of commercial taste may indeed be capable of preventing the form of things from changing too much too quickly, but there are no unalterable forms and many undeniable failures. Whether detected by manufacturer, independent inventor, or consumer, the failure of something to be as light or heavy, thin or thick, or inexpensive or extravagant as a competing or imagined product will institute changes that will ultimately affect in whatever small way the shape of the made world about us. Thomas Edison, whose record 1,093 patents led to some of the most pervasive forms among the artifacts of modern life, was himself caught up in the cycle of technological change that is inescapable. Edison preferred the cylindrical form for sound recordings; indeed, it could be defended as following almost organically from the rotary device that was the first phonograph. When his competitors came out with the flat-disc record, which required a turntable and which would eventually prove to distort sound as the pickup arm progressed from the outer to the inner grooves of the record, Edison at first rejected its form. But when consumers came to prefer the disc because it could be stored more compactly, Edison, who was in the manufacturing business in no small way, did his competitors one better by developing the two-sided record, thus making storage even more efficient. He was not content with things when he saw their shortcomings. As he once wrote in his diary, "Restlessness is discontent—and discontent is the first necessity of progress. Show me a thoroughly satisfied man—and I will show you a failure."

The vast number of things that exist in the world today ensures that there will be ever more tomorrow, for virtually every existing thing is fair game to come under the scrutiny of someone restless and discontented who does not think "well enough" is sufficiently free of faults. The reactionary call to leave well enough alone is a futile one, for the advancement of civilization itself has been a history of the successive correction (and sometimes the overcorrection) of error and fault and failure.

What is well enough for one person may not be so for another, of course. Left-handers have had to learn to live in a world in which door handles, school desks, books, corkscrews, and countless common objects are biased against them. Lefties have to wear borrowed baseball gloves on the wrong hand if their own gloves are at home. But besides fielder's mitts, and the rare school desk, few alternatives to right-handed artifacts have been even remotely available to left-handers, who have simply learned how to live in a right-handed world. Nor do they seem as a rule to express any pressing need for special left-handed devices.

But, as we have seen, specialized artifacts evolve not out of grass-roots needs but out of the idiosyncratic observation of shortcomings in existing things. Thus, inventors and manufacturers have devised left-handed objects, and shops like Anything Left Handed Limited, in London's Brewer Street, offer these in strikingly disorienting catalogues whose pages open from left to right and are numbered accordingly. Though some of the objects offered, like clocks that run counterclockwise, provide more fun than convenience, left-handed garden shears and ladles must seem a godsend. A similar shop exists in San Francisco, where an acquaintance's wife found a left-handed Swiss army knife for him. Not knowing such a thing existed, he explained how he had long gotten by with a conventional model, but now he was anxious to demonstrate how the blades of his new knife could be opened with the fingers of his left hand, and how the corkscrew twisted in the opposite direction from the usual.

Anything Left Handed's kitchen knives have handles shaped to fit the left hand and blades serrated accordingly. Similarly serrated table knives are also available, as are matching pastry forks with the cutting tine on the side where a left-hander needs it. Each item in Anything Left Handed corrects a problem or annoyance that lefties have found in using something designed, whether deliberately or inadvertently, for right-handers. This is a model for the way all artifacts diversify and technology evolves, for as things are used, they reveal their shortcomings—at least to some of us. While inventors, designers, and engineers may not always be the first to see the problems with technology and its objects, they are the ones who do come up with solutions. In the meantime, we tend to accept that ours is a technologically imperfect world and live with its minor annoyances. We may even modify our behavior to accommodate the technology, as left-handers have in adapting to right-handed utensils—until we discover an altered artifact to marvel at and use.

NOTES, BIBLIOGRAPHY, LIST OF ILLUSTRATIONS, AND INDEX

Notes

Full references are given in the bibliography. Short quotations not specifically referenced are from the same sources as referenced quotes nearby in the text.

1 HOW THE FORK GOT ITS TINES

For general background on eating habits and utensils, see especially Bailey, Giblin, Himsworth, and Singleton.

p. 4 "all the tools": Eco and Zorzoli, p. 11.

"any new thing": Basalla, *Evolution,* p. 45.

5 "scramasax": Himsworth, pp. 41–42.

8 "flesh-fork": Beckmann, vol. II, p. 408.

"that they were only used": Bailey, p. 5.

9 "I observed a custom": quoted in Beckmann, vol. II, pp. 412–13n.

"Furcifer": ibid., p. 412.

"an effeminate": ibid., p. 413n.

Ben Jonson: quoted in ibid., p. 413n.

11 Erasmus's 1530 book: quoted in Giblin, pp. 31–32.

12 French book of advice: ibid., pp. 24–25.

Cardinal Richelieu's disgust: ibid., p. 52.

15 The spoon: see Singleton, pp. 4–5; cf. Hume, pp. 180–84.

16 "knives, spoons": Dow, p. 34.

17 "spike and spon": a lunchtime conversation with I. B. Holley, Jr., introduced me to this connection and provided early encouragement for my pursuit of form in the knife and fork.

According to Deetz: Deetz, p. 123. Cf. Furnas, p. 903; Williams, p. 40.

"split spoon": Hooker, p. 97.

"zigzagging": Post, 1945 edition, p. 483; see also illustrations between pp. 448 and 449.

"etiquette manuals": Kasson, p. 44.

"many persons hold": quoted in Williams, p. 40.

p. 18 Frances Trollope: quoted in Turner, p. 58.

"very dirty": Anthony Trollope, quoted in ibid.

Charles Dickens: quoted by John F. Kasson, in Grover, p. 125.

"eating soup": Kasson, in ibid., p. 125.

"everything with it": Kasson, in ibid., p. 125.

Jacob Bronowski: quoted by Ralph Caplan, in *ID,* November-December 1990, p. 11.

some Eskimos: Giblin, pp. 2–6.

19 chopsticks developed: see Giblin, Kleiman, Debra Weiner.

"honorable and upright man": quoted in Giblin, p. 34.

2 FORM FOLLOWS FAILURE

p. 23 "30,000": Norman, pp. 11–12; cf. Biederman, p. 127.

"diversity of things": Basalla, *Evolution,* p. 1.

new chemical substances: *Technology Review,* July 1990, p. 80.

24 "The variety of made things": Basalla, *Evolution,* p. 2.

"the desire of designers": Forty, p. 91.

25 "found in a free": Giedion, p. 396; quoted in Forty, p. 91.

"it is most unlikely": Forty, pp. 92–93.

"Could Montgomery Ward's": ibid., p. 93.

26 "aphorism": ibid., p. 12.

"doctrine": Pye, *Nature and Aesthetics,* pp. 11–12.

"The concept of function": ibid., p. 13.

27 "All designs": ibid., p. 70.

28 "If our metal face": Alexander, p. 19.

29 "Suppose we are given": ibid., p. 23.

30 "Even the most aimless": ibid., pp. 52–53.

31 "although only a few": quoted in ibid., p. 203n.

"the fundamental unit": Basalla, *Evolution,* p. vii.

3 INVENTORS AS CRITICS

p. 34 Jacob Rabinovich: see Brown, pp. 183–85; Rabinow, p. 18.

36 "Inventors are people": Rabinow, p. 212.

"When I see something": in Brown, p. 185.

"continually study the designs": Kamm, p. 142.

Inventors at Work: Brown.

"by far the most": quoted in Holzman, p. 10.

p. 37 "A much better bicycle": quoted in ibid., p. 17.
"terrible-looking bottle": in Brown, p. 370.
"failures and the knowledge": in ibid., p. 368.

38 replacing glass bottles: see Brown, pp. 353–54; cf. *New York Times,* July 7, 1990, Wyeth obituary.
"They tend": in Brown, pp. 77–79.

39 "I think": in ibid., p. 146.

40 "Tools!": Laughlin, p. 36.

41 tungsten-carbide particles: *Nation's Business,* February 1991, p. 72.
"The slot is traditional": Rabinow, p. 195.

42 "If you make": ibid., p. 196.
"When hexagon nuts": Pye, p. 142.

43 Erector sets: Heimberger, pp. 126ff., with thanks to William Petroski.
Meccano sets: Harris, p. 23.
"So long as there are": Coppersmith and Lynx, p. 9.

44 Edwin Land: *New York Times,* March 2, 1991, obituary.
"Invention finds": Usher, p. 11.

45 "I believe that the most": Brunel, p. 492.
bronze powder: Bessemer, pp. 53ff.
"provided the funds": ibid., p. 81.

46 "two-step procedure": Pressman, p. 26.
"This can often": ibid., p. 28.
"You should not spend": ibid., p. 49.
"I knew the art": Rabinow, p. 75.
Rabinow's tuner was patented: U.S. Patent No. 3,119,273.
"No one uses": Rabinow, p. 76.

48 "Discuss how": Pressman, pp. 142–43.

49 "Genius?": quoted in ibid., p. 31.
disagreement: cf. Brown, p. 6; Garrett, p. 4.
"The love of improvement": Bessemer, p. 10.

4 FROM PINS TO PAPER CLIPS

John Eubank, a Detroit pencil collector, and Howard Sufrin, a Pittsburgh collector of antique office supplies (on display at the offices of Premier Business Products), provided hard-to-come-by artifacts and information for this chapter. Betsy Burstein of the

Smithsonian Institution's National Museum of American History kindly provided information about the institution's "National Paper Clip Collection" and some articles from its files.

p. 51 survey: photocopies from Howard Sufrin files.

53 "One man draws": Smith, vol. I, p. 6.

"One fuses metal": quoted in Lubar, p. 258.

sixty feet per minute: Diderot, vol. I, text to plate 185; cf. Greeley, pp. 1286–88.

54 "we must not": Lubar, p. 257.

"mechanical scheming": Howe, quoted in ibid., p. 260.

John Ireland Howe: *Dictionary of American Biography,* vol. IX.

55 "pin money": see Panati, p. 313.

"paper of pins": see Lubar, p. 271.

"bank pins": Morris, p. 12.

57 "These pins": Noesting, p. 7.

sack of emery grit: Armistead, p. 91.

58 "The corners": U.S. Patent No. 43,435.

"as in fasteners similar": quoted in Morris, p. 13.

"effectually secures": U.S. Patent No. 361,439.

59 Latin anagram: see, e.g., Love, p. 2.

60 "fastened paper clips": Segelcke, p. 61.

"It consists of": U.S. Patent No. 675,761.

62 "I am aware": U.S. Patent No. 601,384.

Cornelius Brosnan: U.S. Patent No. 648,841.

63 "first successful bent wire": Anonymous, p. [2].

"This invention relates": U.S. Patent No. 648,841.

66 "of novel shape": U.S. Patent No. 779,522.

67 *"An eye for business":* quoted in Anonymous, p. [3].

"machine for making paper clips:" U.S. Patent No. 636,272.

Brosnan's 1900 patent: U.S. Patent No. 648,841.

"direct ancestor": Anonymous, p. [2].

68 "merely their protector": Morris, p. 12.

"overwhelmingly successful": ibid., p. 13.

69 "derived from the original": Acco International, Inc., "History of the Paper Clip." [Photocopied information sheets.]

"slide on": Army and Navy, p. 349.

"most popular clip": Morris, p. 13.

"If all that survives": Edwards, text on paper clip.

"Could there possibly be": Goldberger, pp. 287–88.

70 "An object": U.S. Patent No. 1,985,866.

75 "We average ten letters": quoted in Richmond (Va.), *Times-Dispatch,* January 20, 1962, p. 2.

76 introduced in the 1950s: see Robert H. Metz, "Tiny Paper Clips Are Big Business," *New York Times,* July 20, 1958.

5 LITTLE THINGS CAN MEAN A LOT

The story of the Minnesota Mining and Manufacturing Company and its products is in the company history, *Our Story So Far,* and in various news releases. Background on staplers comes from material supplied by Stanley Bostitch, Inc.

p. 82 "It lacked proper balance": "Scotch Transparent Tape Celebrates 50th Birthday," 3M news release, [1980].

84 Art Fry: 3M news releases; cf. Time-Life, p. 75.

"some of the hymnal pages": "Stick-to-it-spirit Takes the Post-it Brand Note from Brainstorm to Marketplace," 3M news release, June 18, 1987.

"bootlegging": ibid.; cf. Rabinow, p. 34.

"temporarily permanent": Time-Life, p. 75.

85 "who had to accept": "3M Researcher's Yen for Hymnal Marker Produces One of the 1980s Top Selling Products," 3M marketing services note, June 18, 1987.

90 "the use of staples": Bostitch, "A Young Company with Half a Century of Experience," p. 5.

91 "new models": ibid.

6 STICK BEFORE ZIP

The story of the zipper is told in the Talon company history, abridged from a manuscript by James Gray. An independent treatment appears in the article by Federico.

p. 93 Thomas Woodward: U.S. Patent No. 2,609.

94 Walter Hunt: U.S. Patent No. 6,281.

assigned to the draftsman: see, e.g., de Bono, p. 123.

p. 96 buttons on men's garments: see Feldman, pp. 237–38.
98 Elias Howe: Lewis Weiner, p. 132; cf. U.S. Patent No. 8540.
Whitcomb Judson: Gray; cf. Federico; see also Lewis Weiner.
one of Judson's first: U.S. Patent No. 504,037.
A second patent: U.S. Patent No. 504,438.
100 "perfect the details": Gray, p. 21.
102 "A pull and it's done!": advertisement reproduced in ibid., p. 24.
"A leaflet": ibid., p. 26.
103 "the hook and eye principle": ibid., p. 25.
Otto Frederick Gideon Sundback: ibid., pp. 28ff.
104 "to keep Judson's machine running": ibid., p. 23.
"His shrewd eye": ibid. pp. 29–30.
"fully saturated": quoted in ibid., p. 30.
James O'Neill: ibid., pp. 30, 32.
105 "To one side": ibid., p. 33.
"hidden hook": U.S. Patent No. 1,236,784.
"It doesn't seem to me": quoted in Gray, p. 34.
106 "An obscure company": ibid., p. 37.
"Cross over quickly": quoted in ibid., p. 34.
Hookless No. 2: ibid., pp. 38ff.
Scientific American: June 1983; see Lewis Weiner.
108 "First, a demand": Gray, p. 40.
a host of others: ibid., pp. 39–40; Federico, pp. 862ff.
109 orders for the hookless: Gray, pp. 42–45.
110 S-L machine: ibid., pp. 52–53.
111 Bertram G. Work: ibid., p. 60.
"Talon": ibid., pp. 71, 85.

7 TOOLS MAKE TOOLS

p. 115 "One stool was called": Sturt, *William Smith,* pp. 71–72.
116 "ribber": ibid., p. 73.
"A stranger entering": in Jackson and Day, p. 7.
117 "an iron tool": Agricola, p. 269.
119 earliest metal saws: see Jackson and Day, p. 74.
tooth-embedded jawbones: see Beckmann, vol. I, pp. 223–24.
121 "It's a rare music": Underhill, *Woodwright's Companion,* p. 165.

p. 123 "Laborious it was": Sturt, *Wheelwright's Shop,* pp. 35, 39–40.

125 duplex saw: *Chronicle* (Early American Industries Association), March 1989, p. 13.

126 "Kentucky, Ohio": see Basalla, *Evolution,* p. 89.

128 *The Hammer:* Baird and Comerford.

129 "a startling or unique": Pressman, p. 77.

"The best tack hammer": quoted in *Chronicle* (Early American Industries Association), March 1989, p. 13.

8 PATTERNS OF PROLIFERATION

p. 131 "An early list": MacLachlan, p. 33.

"Ideal Olive Fork": Turner, p. 217.

132 "A number of factors": MacLachlan, p. 25.

"Tines on early models": ibid., p. 9.

134 "By 1898": Rainwater, in Grover, pp. 181–82.

Herbert Hoover: Turner, p. 185.

136 "One of the fears": Post, pp. 631–32.

138 twenty-five years: Turner, p. 54; cf. Sears, Roebuck *Catalogue*, Fall and Winter, 1928–29, pp. 744–51, for ranges of quality available in the 1920s.

139 "In selecting her silver": Post, pp. 626–27.

140 "The small fork": ibid., p. 629.

142 "always with the primary object": Coppersmith and Lynx, p. 20.

"Having invited his friends": ibid., pp. 22–23.

"The menu": ibid., p. 20.

144 "Seven and even nine": Hall, pp. 77–78; cf. Grover, Kasson, Williams.

"for a small dinner": Hall, p. 80.

"Short dinners": Learned, p. 85.

145 "In order to give": Hall, pp. 80–81.

146 "All English-speaking nations": ibid., pp. 86–87.

147 "In England and her colonies": Bradley, pp. 189–90.

"made dishes": A Member, *Manners and Tone,* p. 93.

"much more difficult": Bradley, p. 190.

"It is an affectation": quoted in Williams, p. 42.

148 "cutting fork": Turner, p. 180.

"with a fork": quoted in Williams, p. 42.

p. 148 "wasteful": Post, p. 629.

149 "It was then discovered": A Member, *Manners and Rules,* p. 118.

"at all formal dinners": Bradley, p. 180.

"was so very inconvenient": Hall, p. 85.

150 "fretwork trimmings": Post, p. 629.

stainless-steel knife blades: Himsworth, p. 74.

151 "the ten most essential": Rainwater, in Grover, p. 202.

9 DOMESTIC FASHION AND INDUSTRIAL DESIGN

p. 157 "possible only to the longest purse": Post, p. 627.

158 "very best English goods": Army & Navy.

"on bad silver": Post, p. 627.

159 "quality" and "charm": Sears, Roebuck *Catalogue*, Fall and Winter 1928–29, pp. 744–45.

160 handbook of hammers: Baird and Comerford.

country craft tools: illustration reproduced in Basalla, *Evolution,* pp. 4–5.

"potters could not afford": quoted in Forty, p. 18.

161 "style consists": Viollet-le-Duc, p. 175.

162 "His first care": ibid., p. 177.

"But the coppersmiths": ibid., p. 178.

164 "physical appearance": Loewy, *Never Leave,* p. 66.

165 "Once in a while": ibid., pp. 74–75.

"shocked by the fact": Loewy, *Industrial Design,* p. 51.

166 "success finally came": ibid., p. 52.

Sigmund Gestetner: ibid., p. 60.

167 "Your present models": Loewy, *Never Leave,* p. 189.

sewing needle: ibid., p. 195.

Lucky Strike: ibid., pp. 146–49.

168 railroad commission: ibid., pp. 135–41.

169 "no manufacturer": ibid., p. 187.

"Walk through": Dreyfuss, p. 203.

five-point formula: ibid., p. 160.

170 MAYA: Loewy, *Never Leave,* p. 278.

"survival form": Dreyfuss, pp. 57–58.

"strive for a delicate balance": Heskett, pp. 177–78.

10 THE POWER OF PRECEDENT

p. 171 "It became": Jewitt, pp. 52–53.

172 mottoes and verses: quoted in Jewitt, pp. 53–54.

174 insightful article: Ferguson, "Mind's Eye."

175 "The Ariel": Heskett, pp. 177–78.

176 "radical innovation": *ID,* May-June 1990, p. 72.

"Now the *important*": Loewy, *Never Leave,* p. 313.

177 "Changes are": ibid., p. 314.

"a clear-cut case": ibid., p. 358.

178 "I unfolded my easel": ibid., p. 359.

"Most of my work": quoted in Wolff, pp. 44–45.

179 "The following": Pressman, p. 159.

"other tricks": ibid., p. 174.

180 "A claim that is short": ibid., pp. 174–75.

"In a flying machine": U.S. Patent No. 821,393.

181 "craft in the tradition": Brown, p. 1.

182 Sydney Opera House: see, e.g., *ENR,* May 17, 1990, p. 26.

183 "It is relatively easy": Billington, "Aesthetics," p. 11.

11 CLOSURE BEFORE OPENING

For general background on the tin can, see especially Church. For the development of the aluminum beverage can, see especially various articles in *Modern Metals,* as noted below.

p. 185 A prize: Church, p. 22.

"tin canister": de Bono, p. 110.

186 "Cut round": ibid., p. 113.

"sometimes heavier": Panati, p. 115.

one pound empty: de Bono, p. 113.

"the first can-openers": ibid.

187 "part bayonet": Panati, p. 115.

188 "The advantages": U.S. Patent No. 19,063.

Bull's Head: see de Bono, p. 113.

189 William Lyman: U.S. Patent No. 105,346.

familiar style of wheeled opener: see Panati, p. 116.

Sears, Roebuck catalogue: Fall and Winter, 1928–29, p. 801.

p. 191 fortified wine: *New York Times,* January 31, 1991, p. C18.

194 church key is a simple lever: see Edwards, "church key and bottle opener" page.

197 Kaiser Aluminum: *Modern Metals,* January 1972, pp. 65, 67.

Adolph Coors Company: ibid., February 1959, pp. 62–63; January 1972, pp. 64ff.

Reynolds Metals and Alcoa: ibid., August 1967, p. 57.

198 can top . . . must be thicker: see ibid., December 1979, p. 27.

"There must be": Dayton *Daily News,* October 27, 1989, pp. 1A, 6A.

199 "I personally did not": Time-Life, p. 74. Cf., e.g., U.S. Patents Nos. 2,153,344; 2,978,140; 3,059,808.

201 to solve the loose-tab problem: see, e.g., *Machine Design,* November 25, 1976, p. 8.

"easy open ecology end": see U.S. Patent No. 3,877,604.

"Since most people": U.S. Patent No. 3,870,001.

203 Francis Silver: U.S. Patent No. 3,877,606.

Royal Crown: *Modern Metals,* July 1964, p. 86.

Coke and Pepsi: ibid., August 1967, p. 57.

recycling aluminum cans: ibid., January 1972, pp. 72, 74; May 1989, pp. 76ff. Cf. *Resource Recycling,* October 1990, pp. 16, 18–19.

204 records on recycling: *Modern Metals,* May 1989, p. 76.

liquid nitrogen: *Iron Age,* November 1988, p. 34.

steel-can industry: *Scientific American,* February 1989, pp. 72–73.

205 "While container reclosure": U.S. Patent No. 4,673,099.

207 "It is believed": U.S. Patent No. 4,951,835.

12 BIG BUCKS FROM SMALL CHANGE

p. 209 "remarkable things occur": Aristotle, p. 331.

"Civil engineering is the art": in J. G. Watson, p. 9; cf. Garth Watson, p. 19.

210 "Civil engineering is the profession": American Society of Civil Engineers, *Official Register 1992,* p. 293.

"Why do they make beds": Aristotle, p. 395.

211 "They do not cord them": ibid., pp. 395–96.

olive tree: Homer, p. 257.

212 American rope beds: Graves, p. 59.

213 "It would be well": quoted in *Engineering Education,* July-August 1990, p. 524.

p. 216 "Ideally in a patent search": Edelson, pp. 97–98.

217 computerize the files: *Design News,* November 5, 1990, pp. 96–97.

218 Robert Kearns: Durham, North Carolina, *Morning Herald,* November 18, 1990, p. A8.

13 WHEN GOOD IS BETTER THAN BEST

p. 220 McDonald's: *New York Times,* November 1 and 2, 1990; *Modern Plastics,* October 1987, p. 15; September 1990, p. 53; December 1990, pp. 42–45, 49.

224 "polystyrene production process": *New York Times,* November 1, 1990, p. C17.

"Burger King applauds": *New York Times,* November 7, 1990, advertisement.

225 "improvements in cooking": *New York Times,* November 2, 1990, p. C5.

Chinese wheelbarrow: Mayne.

231 "an ingenious example": Caplan, pp. 160–61; see also *Industrial Design,* July–August 1984, p. 8.

14 ALWAYS ROOM FOR IMPROVEMENT

p. 237 Russell Baker: *New York Times,* November 13, 1990, op-ed page.

"new telephone systems": Norman, p. vii.

"count upon finding": ibid., p. 6.

240 *Design News:* May 21, 1990, editorial; October 22, 1990, pp. 130–32, 135.

241 Goldstar Electronics: see *New York Times,* November 6, 1990, advertising column.

242 "usable design": Norman, pp. viii, ix.

245 "This is how sellers": Aristotle, p. 347.

248 Jacob Rabinow: Rabinow, pp. 223–24.

249 "Restlessness is discontent": quoted in Jones, p. 2.

Bibliography

Agricola, Georgius. *De Re Metallica.* Translated by Herbert Clark Hoover and Lou Henry Hoover. New York: Dover Publications, 1950.

Alexander, Christopher. *Notes on the Synthesis of Form.* Cambridge, Mass.: Harvard University Press, 1964.

Anonymous. "Behold the Lowly Paper Clip . . . It's Still a 'Gem.'" *Office Products,* October 1975.

Aristotle. *Minor Works.* Translated by W. S. Hett. Cambridge, Mass.: Harvard University Press, 1980.

Armistead, Don. "The Lore of the Abrasive Little Strawberry," *The Chronicle of the Early American Industries Association,* September 1991, pp. 91–92.

Army & Navy Co-operative Society. *The Very Best English Goods: A Facsimile of the Original Catalogue of Edwardian Fashions, Furnishings, and Notions Sold . . . in 1907.* New York: Frederick A. Praeger, 1969.

Bacon, Francis. *The Advancement of Learning and New Atlantis.* Edited by Arthur Johnston. Oxford: Clarendon Press, 1974.

Bailey, C. T. P. *Knives and Forks.* London: The Medici Society, 1927.

Baird, Ron, and Comerford, Dan. *The Hammer: The King of Tools.* Privately printed, 1989.

Barsley, Michael. *The Left-handed Book: An Investigation into the Sinister History of Left-handedness.* London: Souvenir Press, 1966.

Basalla, George. *The Evolution of Technology.* Cambridge: University Press, 1988.

———. "Transformed Utilitarian Objects." *Winterthur Portfolio 17* (Winter 1982): 183–201.

Beckmann, John. *A History of Inventions, Discoveries, and Origins.* Translated by William Johnston. 4th ed., revised and enlarged by William Francis and J. W. Griffith. London: Henry G. Bohn, 1846.

Benker, Gertrud. *Das Wilkens-Brevier vom silbernen Besteck: Wissenswertes von A–Z.* Bremen: M. H. Wilkens & Sohne GmbH, [1990].

Bessemer, Henry. *Sir Henry Bessemer, F.R.S.: An Autobiography.* London: Offices of *Engineering,* 1905.

Biederman, Irving. "Recognition-by-Components: A Theory of Human Image Understanding." *Psychological Review* 94 (1987): 115–47.

Bijker, Wiebe E., Hughes, Thomas P., and Pinch, Trevor, eds. *The Social Construction of Technological Systems: New Directions in the Sociology and History of Technology.* Cambridge, Mass.: MIT Press, 1987.

Billington, David P. "Aesthetics in Bridge Design—The Challenge." In *Bridge Design: Aesthetics and Developing Technologies.* Edited by Adele Fleet Bacow and Kenneth E. Kruckemeyer. Boston: Massachusetts Department of Public Works and Massachusetts Council on the Arts and Humanities, 1986, pp. 3–16.

———. *The Tower and the Bridge: The New Art of Structural Engineering.* New York: Basic Books, 1983.

Boggs, Robert N. "Rogues' Gallery of 'Aggravating Products.' " *Design News,* October 22, 1990, pp. 130–33.

Bradley, Mrs. Julia M. *Modern Manners and Social Forms.* Chicago: James B. Smiley, 1889.

Bronowski, Jacob. *The Origins of Knowledge and Imagination.* New Haven, Conn.: Yale University Press, 1978.

Brown, Kenneth A. *Inventors at Work: Interviews with 16 Notable American Inventors.* Redmond, Wash.: Tempus Books, 1988.

Brunel, Isambard. *The Life of Isambard Kingdom Brunel, Civil Engineer.* London: Longmans, Green, 1870.

Burlingame, Roger. *Inventors Behind the Inventor.* New York: Harcourt, Brace, and Company, 1947.

Butterworth, Benjamin. *The Growth of Industrial Art.* Washington, D.C.: Government Printing Office, 1888.

Caplan, Ralph. *By Design: Why There Are No Locks on the Bathroom Doors in the Hotel Louis XIV and Other Object Lessons.* New York: St. Martin's Press, 1982.

The Chronicle of The Early American Industries Association, various issues.

Church, Fred L. "The Tin Can: After 190 Years, Still Going Strong." *Modern Metals,* February 1991, pp. 22, 24, 26, 28, 30, 32.

Clarke, Donald, ed. *The Encyclopedia of Inventions.* New York: Galahad Books, 1977.

Coppersmith, Fred., and Lynx, J. J. *Patent Applied For: A Century of Fantastic Inventions.* [London]: Co-ordination Ltd., 1949.

Couch, Tom D. *The Bishop's Boys: A Life of Wilbur and Orville Wright.* New York: W. W. Norton, 1989.

[Day, C. W.] *Hints on Etiquette: And the Usages of Society with a Glance at Bad Habits.* New York: E. P. Dutton, 1951.

de Bono, Edward. *Eureka!: An Illustrated History of Inventions from the Wheel to the Computer.* New York: Holt, Rinehart and Winston, 1974.

Deetz, James. *In Small Things Forgotten: The Archaeology of Early American Life.* Garden City, N.Y.: Anchor Press/Doubleday, 1977.

de Vries, Leonard. *Victorian Inventions.* New York: American Heritage Press, 1971.

Diderot, Denis. *A Diderot Pictorial Encyclopedia of Trades and Industry . . .* Edited by Charles Coulston Gillispie. New York: Dover Publications, 1959.

Dow, George Francis. *Every Day Life in the Massachusetts Bay Colony.* Boston: Society for the Preservation of New England Antiquities, 1935.

Dreyfuss, Henry. *Designing for People.* New York: Paragraphic Books, 1967.

Eco, Umberto, and Zorzoli, G. B. *The Picture History of Inventions: From Plough to Polaris.* Translated by Anthony Lawrence. New York: Macmillan, 1963.

Edelson, Nathan. "An Inventor Goes to Washington." *Design News,* November 5, 1990, pp. 95–99.

Edison, Thomas Alva. *The Diary and Sundry Observations.* Edited by Dagobert D. Runes. New York: Philosophical Library, 1948.

Edison Lamp Works. *Pictorial History of the Edison Lamp.* Harrison, N.J.: General Electric Company, [ca. 1915].

Edwards, Owen. *Elegant Solutions: Quintessential Technology for a User-friendly World.* New York: Crown, 1989.

Farrell, Christopher J. "A Theory of Technological Progress." Unpublished manuscript.

Federico, P. J. "The Invention and Introduction of the Zipper." *Journal of the Patent Office Society* 28 (December 1946): 855–76.

Feldman, David. *Why Do Clocks Run Clockwise? and Other Imponderables: Mysteries of Everyday Life.* New York: Harper & Row, 1987.

Ferguson, Eugene S. "The Mind's Eye: Nonverbal Thought in Technology." *Science* 197 (August 26, 1977): 827–36.

———. *Engineering and the Mind's Eye.* Cambridge, Mass.: MIT Press, 1992.

Forty, Adrian. *Objects of Desire.* New York: Pantheon, 1986.

Friedel, Robert. *A Material World: An Exhibition at the National Museum of American History.* Washington, D.C.: Smithsonian Institution, 1988.

Furnas, J. C. *The Americans: A Social History of the United States.* New York: Putnam's, 1969.

Garrett, Alfred B. *Flash of Genius.* Princeton, N.J.: Van Nostrand, 1963.

Giblin, James Cross. *From Hand to Mouth: Or, How We Invented Knives, Forks, Spoons, and Chopsticks & the Table Manners to Go With Them.* New York: Crowell, 1987.

Giedion, Siegfried. *Mechanization Takes Command: A Contribution to Anonymous History.* New York: W. W. Norton, 1969.

Glegg, Gordon L. *The Development of Design.* Cambridge: University Press, 1981.

Goldberger, Paul. *On the Rise: Architecture and Design in a Postmodern Age.* New York: Penguin Books, 1985.

Goodman, W. L. *The History of Woodworking Tools.* New York: David McKay, 1964.

Graves, Donald. "Bedding, Beds, and Bedsteads." *Early American Life,* October 1987, pp. 56–59, 72.

[Gray, James.] *Talon, Inc.: A Romance of Achievement.* Meadville, Pa.: Talon, Inc., 1963.

Greeley, Horace, et al. *The Great Industries of the United States: Being an Historical Summary of the Origin, Growth, and Perfection of the Chief Industrial Arts of this Country.* Hartford, Conn.: J. B. Burr & Hyde, 1873.

Grover, Kathryn, ed. *Dining in America: 1850–1900.* Amherst: University of Massachusetts Press, 1987.

Gurcke, Karl. *Bricks and Brickmaking: A Handbook for Historical Archaeology.* Moscow, Idaho: University of Idaho Press, 1987.

Hagan, Tere. *Silverplated Flatware: An Identification and Value Guide.* Rev. 4th ed. Paducah, Ky.: Collector Books, 1990.

Hall, Florence Howe. *Social Customs.* Boston: Estes and Lauriat, 1887.

Harris, Alan. "Model Childhood," *New Civil Engineer,* June 13, 1991, p. 23.

Harter, R. J. "Patent It Yourself," *Design News,* November 18, 1991, pp. 93–97.

Heimburger, Donald J., ed. *A. C. Gilbert's Heritage.* River Forest, Ill.: Heimburger House, 1983.

Heskett, John. *Industrial Design.* New York: Oxford University Press, 1980.

Himsworth, J. B. *The Story of Cutlery: From Flint to Stainless Steel.* London: Ernest Benn, 1953.

Hindle, Brooke. *Technology in America: Needs and Opportunities for Study.* With a directory of artifact collections, by Lucius F. Ellsworth. Chapel Hill: University of North Carolina Press, 1966.

Hindle, Brooke, and Lubar, Steven. *Engines of Change: The American Industrial Revolution, 1790–1860.* Washington, D.C.: Smithsonian Institution Press, 1986.

Holzman, David. "Masterful Tinkering of Genius." *Insight,* June 25, 1990, pp. 8–17.

Homer. *The Odyssey: The Story of Odysseus.* Translated by W. H. D. Rouse. New York: New American Library, 1949.

Hooker, Richard J. *Food and Drink in America: A History.* Indianapolis: Bobbs-Merrill, 1981.

Hounshell, David A. *From the American System to Mass Production, 1800–1932: The Development of Manufacturing Technology in the United States.* Baltimore: Johns Hopkins University Press, 1984.

Hughes, Thomas P. *American Genesis: A Century of Invention and Technological Enthusiasm, 1870–1970.* New York: Viking, 1989.

Hume, Ivor Noël. *A Guide to Artifacts of Colonial America.* New York: Alfred A. Knopf, 1970.

International Paper Company. *Pocket Pal: A Graphic Arts Digest for Printers and Advertising Production Managers.* New York: International Paper Company, 1966.

Jackson, Albert, and Day, David. *Tools and How to Use Them: An Illustrated Encyclopedia.* New York: Alfred A. Knopf, 1978.

Jenkins, J. Geraint. *The English Farm Wagon: Origins and Structure.* Lingfield, Surrey, Eng.: Oakwood Press, 1961.

Jewitt, Llewellynn. *The Wedgwoods: Being a Life of Josiah Wedgwood; with Notices of His Works and Their Productions, Memoirs of the Wedgwood and Other Families, and a History of the Early Potteries of Staffordshire.* London: Virtue Brothers, 1865.

Jones, Nancy Cela, ed. *Edison and His Invention Factory: A Photo Essay.* [Washington, D.C.]: Eastern National Park and Monument Association, 1989.

Kamm, Lawrence J. *Successful Engineering: A Guide to Achieving Your Career Goals.* New York: McGraw-Hill, 1989.

Kasson, John F. *Rudeness and Civility: Manners in Nineteenth-Century Urban America.* New York: Hill and Wang, 1990.

Kleiman, Dena. "Older Than Forks, Safer Than Knives." *New York Times,* January 17, 1990, p. C4.

Klenck, Thomas. "Pliers." *Popular Mechanics,* September 1989, pp. 71–74.

Laughlin, M. Penn. *Money from Ideas: A Primer on Inventions and Patents.* Chicago: Popular Mechanics Press, 1950.

Learned, Mrs. Frank. *The Etiquette of New York To-day.* New York: Frederick A. Stokes Co., 1906.

Loewy, Raymond. *Industrial Design.* Woodstock, N.Y.: Overlook Press, 1979.

———. *Never Leave Well Enough Alone.* New York: Simon and Schuster, 1951.

Love, A. E. H. *A Treatise on the Mathematical Theory of Elasticity.* New York: Dover Publications, n.d. [reprint of 4th ed., 1927.]

Lubar, Steven. "Culture and Technological Design in the 19th-Century Pin Industry: John Howe and the Howe Manufacturing Company." *Technology and Culture* 28 (April 1987): 253–82.

MacLachlan, Suzanne. *A Collectors' Handbook for Grape Nuts.* Privately printed, 1971.

MacLeod, Christine. *Inventing the Industrial Revolution: The English Patent System, 1660–1800.* Cambridge: University Press, 1988.

Mason, Otis T. *The Origins of Invention: A Study of Industry Among Primitive Peoples.* London: Walter Scott, 1907.

Mayne, Charles. "Note on the Chinese Wheelbarrow." *Minutes of Proceedings of the Institution of Civil Engineers* 127 (1897): 312–14.

McClure, J. B., ed. *Edison and His Inventions . . .* Chicago: Rhodes & McClure, 1879.

A Member of the Aristocracy. *Manners and Rules of Good Society: Or Solecisms to Be Avoided.* 33rd ed. London: Frederick Warne and Co., 1911.

———. *Manners and Tone of Good Society: Or Solecisms to Be Avoided.* 4th ed. London: Frederick Warne and Co., n.d.

Mercer, Henry C. *Ancient Carpenters' Tools: Together with Lumbermen's, Joiners' and Cabinet Makers' Tools in Use in the Eighteenth Century.* Doylestown, Pa.: Bucks County Historical Society, 1951.

Minnesota Mining and Manufacturing Company. *Our Story So Far: Notes from the First 75 Years of 3M Company.* St. Paul, Minn.: Minnesota Mining and Manufacturing Company, 1977.

Morris, Danny A. "Emanuel Fritz Paper Clip Collection," *American Collector,* July 1973, pp. 12–13.

Moxon, Joseph. *Mechanick Exercises, or the Doctrine of Handy-Works.* Morristown, N.J.: Astragal Press, 1989 [reprint of 1703 ed.].

Mumford, Lewis. *Technics and Civilization.* New York: Harcourt Brace Jovanovich, 1963.

Noesting, Inc. *Catalogue.* 1989.

Norman, Donald A. *The Design of Everyday Things.* New York: Doubleday, 1989.

Panati, Charles. *Panati's Extraordinary Origins of Everyday Things.* New York: Harper & Row, 1987.

Papanek, Victor. *Design for Human Scale.* New York: Van Nostrand Reinhold, 1983.

———. *Design for the Real World: Human Ecology and Social Change.* New York: Pantheon, 1971.

Park, Robert. *Inventor's Handbook.* White Hall, Va.: Betterway Publications, 1986.

Petroski, Henry. *The Pencil: A History of Design and Circumstance.* New York: Alfred A. Knopf, 1990.

———. *To Engineer Is Human: The Role of Failure in Successful Design.* New York: St. Martin's Press, 1985.

Pinchot, Gifford, III. *Intrapreneuring: Why You Don't Have to Leave the Corporation to Become an Entrepreneur.* New York: Harper & Row, 1985.

Pitt-Rivers, A. Lane-Fox. *The Evolution of Culture and Other Essays.* Edited by J. L. Myers. Oxford: Clarendon Press, 1906.

Post, Emily. *Etiquette: "The Blue Book of Social Usage."* New and enlarged ed. New York: Funk & Wagnalls, 1927. [Also other editions, as noted.]

Pressman, David. *Patent It Yourself.* Berkeley, Calif.: Nolo Press, 1985. [Also 3rd ed., 1991]

Pye, David. *The Nature and Aesthetics of Design.* London: Barrie & Jenkins, 1978. [Reprinted, London: The Herbert Press, 1988.]

———. *The Nature and Art of Workmanship.* Cambridge: University Press, 1968.

Rabinow, Jacob. *Inventing for Fun and Profit.* San Francisco: San Francisco Press, 1990.

Rainwater, Dorothy T. and H Ivan. *American Silverplate.* West Chester, Pa.: Schiffer Publishing, 1988.

Read, Herbert. *Art and Industry: The Principles of Industrial Design.* London: Faber and Faber, 1934.

Richman, Miriam. "Antique Woodworking Tools." *Early American Life,* August 1990, pp. 26–28, 58.

Rossman, Joseph. *The Psychology of the Inventor: A Study of the Patentee.* New and revised ed. Washington, D.C.: Inventors Publishing Company, 1931.

Rybczynski, Witold. *Home: A Short History of an Idea.* New York: Penguin Books, 1987.

Schaefer, Herwin. *Nineteenth Century Modern: The Functional Tradition in Victorian Design.* New York: Praeger, 1970.

Schroeder, Fred E. H. "More 'Small Things Forgotten': Domestic Electrical Plugs and Receptacles, 1881–1931." *Technology and Culture* 27 (July 1986): 525–43.

Sears, Roebuck and Company. *Catalogue.* Various editions.

Segelcke, Nanna. *Made in Norway.* Oslo: Dreyer, 1990.

Simon, Herbert A. *The Sciences of the Artificial.* 2nd ed. Cambridge, Mass.: MIT Press, 1981.

Singleton, H. Raymond. *A Chronology of Cutlery.* Sheffield, Eng.: City Museums, 1970.

Smith, Adam. *An Inquiry into the Nature and Causes of the Wealth of Nations.* Oxford: Clarendon Press, 1880.

Squires, Arthur L. *The Tender Ship: Governmental Management of Technological Change.* Boston: Birkhäuser, 1986.

Steadman, Philip. *The Evolution of Designs: Biological Analogy in Architecture and the Applied Arts.* Cambridge: University Press, 1979.

Straub, Hans. *A History of Civil Engineering: An Outline from Ancient to Modern Times.* Cambridge, Mass.: MIT Press, 1964.

Strung, Norman. *An Encyclopedia of Knives.* Philadelphia: J. B. Lippincott, 1976.

Sturt, George. *The Wheelwright's Shop.* Cambridge: University Press, 1934.

———. *William Smith, Potter and Farmer: 1790–1858.* Firle, Sussex, Eng.: Caliban Books, 1978. [Facsimile of original ed., 1919, published under the pseudonym George Bourne.]

Tannahill, Reay. *Food in History.* New ed. New York: Crown, 1989.

Thompson, D'Arcy Wentworth. *On Growth and Form.* Edited by John Tyler Bonner. Cambridge: University Press, 1961.

Time-Life Books, eds. *Inventive Genius.* Alexandria, Va.: Time-Life Books, 1991.

Timoshenko, Stephen P. *History of Strength of Materials: With a Brief Account of the History of Theory of Elasticity and Theory of Structures.* New York: Dover Publications, 1983. [Reprint of 1953 ed.]

Tunis, Edwin. *Colonial Craftsmen and the Beginnings of American Industry.* Cleveland: World, 1965.

Turner, Noel D. *American Silver Flatware, 1837–1910.* South Brunswick, N.J.: A. S. Barnes, 1972.

Underhill, Roy. *The Woodwright's Companion: Exploring Traditional Woodcraft.* Chapel Hill: University of North Carolina Press, 1983.

———. *The Woodwright's Shop: A Practical Guide to Traditional Woodcraft.* Chapel Hill: University of North Carolina Press, 1981.

Usher, Abbott Payson. *A History of Mechanical Inventions.* New York: McGraw-Hill, 1929.

Vanderbilt, Amy. *Amy Vanderbilt's* New *Complete Book of Etiquette: The Guide to Gracious Living.* Garden City, N.Y.: Doubleday, 1963.

Vincenti, Walter G. *What Engineers Know and How They Know It: Analytical Studies from Aeronautical History.* Baltimore: Johns Hopkins University Press, 1990.

Viollet-le-Duc, Eugène Emmanuel. *Discourses on Architecture.* Translated by Henry van Brunt. Boston: James R. Osgood, 1875.

Vitruvius. *The Ten Books on Architecture.* Translated by Morris Hicky Morgan. New York: Dover Publications, 1960.

Vogue. *Vogue's Book of Etiquette and Good Manners.* New York: Condé Nast, 1969.

Wallace & Sons Manufacturing Company. *How to Set the Table.* 19th ed. Haverhill, Mass.: Horace N. Noyes, [ca. 1915].

Ward, Montgomery, and Company. *Catalogue.* Various editions.

Watson, Garth. *The Civils—The Story of the Institution of Civil Engineers.* London: Thomas Telford, 1988.

Watson, J. G. *A Short History.* London: Institution of Civil Engineers, 1982.

Wedgwood, Josiah. *Selected Letters.* Edited by Ann Finer and George Savage. London: Cory, Adams & Mackay, 1965.

Weiner, Debra. "Chopsticks: Ritual, Lore and Etiquette." *New York Times,* December 26, 1984, p. III, 3.

Weiner, Lewis. "The Slide Fastener." *Scientific American,* June 1983, pp. 132–36, 138, 143–44.

White, Francis Sellon. *A History of Inventions and Discoveries: Alphabetically Arranged.* London: C. and J. Rivington, 1827.

Wilkens Bremer Silberwaren AG. Various catalogues and publications.

Williams, Susan. *Savory Suppers and Fashionable Feasts: Dining in Victorian America.* New York: Pantheon, 1985.

Wolff, Michael F. "Inventing at Breakfast." *IEEE Spectrum,* May 1975, pp. 44–49.

List of Illustrations

Index

Italicized page numbers refer to figures and figure captions.

A NOTE ABOUT THE AUTHOR

Henry Petroski is professor of civil engineering and chairman of the department of civil and environmental engineering at Duke University. His previous books are *The Pencil: A History of Design and Circumstance; To Engineer Is Human: The Role of Failure in Successful Design,* the concept of which he developed into a BBC television documentary; and *Beyond Engineering: Essays and Other Attempts to Figure Without Equations.* He also writes the engineering column in *American Scientist.*

A NOTE ON THE TYPE

This book was set in Caledonia, a typeface designed by W(illiam) A(ddison) Dwiggins (1880–1956) for the Merganthaler Linotype Company in 1939. Dwiggins chose to call his new typeface Caledonia, the Roman name for Scotland, because it was inspired by the Scottish types cast about 1833 by Alexander Wilson & Son, Glasgow type founders. However, there is a calligraphic quality about Caledonia that is totally lacking in the Wilson types.

Dwiggins referred to an even earlier typeface for this "liveliness of action"—one cut around 1790 by William Martin for the printer William Bulmer. Caledonia has more weight than the Martin letters, and the bottom finishing strokes (serifs) of the letters are cut straight across, without brackets, to make sharp angles with the upright stems, thus giving a modern-face appearance.

W. A. Dwiggins began an association with the Merganthaler Linotype Company in 1929 and over the next twenty-seven years designed a number of book types, the most interesting of which are Metro, Electra, Caledonia, Eldorado, and Falcon.

Printed and bound by
Halliday Lithographers
West Hanover, Massachusetts

Designed by Brooke Zimmer